ALLE ZEIT WACH 1842

Moshe R. Heller (Editor)
ASIMUTH

Automotive Simulation '91

Proceedings of the 3rd European Cars/Trucks, Simulation Symposium Schliersee, Germany, October 1991

With 121 Figures

Springer-Verlag
Berlin Heidelberg NewYork
London Paris Tokyo
Hong Kong Barcelona Budapest

Editor:
Moshe R. Heller
ASIMUTH - Applied Simulation Technology GmbH
Planegger Straße 47
8000 München 60
Germany

ISBN-13:978-3-642-84588-8 e-ISBN-13:978-3-642-84586-4
DOI: 10.1007/978-3-642-84586-4

Library of Congress Cataloging-in-Publication Data
Cars and trucks simulation : 3rd European symposium, Schliersee, FRG, October 28-30, 1991 / Moshe R. Heller, editor.
Papers presented at the 3rd European Cars/Trucks Simulation Symposium.
ISBN-13:978-3-642-84588-8

1. Automobiles--Design and construction--Computer simulation--Congresses.
2. Trucks--Design and construction--Computer simulation--Congresses.
I. Heller, Moshe R.
II. European Cars/Trucks Simulation Symposium (3rd : 1991 : Schliersee, Germany)
TL240.C37 1991
629.2'3'0113--dc20 91-30229

Softcover reprint of the hardcover 1st edition 1991

Typesetting: Camera ready by authors

61/3020-543210 - Printed on acid-free paper

Preface

Welcome to Bavaria - Germany - to the THIRD EUROPEAN CARS/TRUCKS SIMULATION SYMPOSIUM. That Schliersee traditional workshop-type meeting is a follow-up to the first and the second Symposia which took place in May 1984 and May 1989 respectively.

The objective of gathering together is to cover most of the aspects of Automotive Mathematical Modelling and Simulation in theory and practice to promote the exchange of knowledge and experience between different national and international research groups in that field, taking into consideration that every seventh German employee is related to the automotive industry. This effect is also in power at least with the traditional Detroit (U.S.A.) Automotive Industries and the growing up Japanease as well.

Futhermore, there is to strenghten the international contact between developers and users of modelling and simulation techniques considering the "new world order" started in 1991 with no borders between West and East affected by the Golf-War and followed up by the "open" European Community borders of 1992.

The traditional International Conference jointly promoted by ASIMUTH - Applied Simulation Technology and some other members of the Society of Computer Simulaton created an interest to publish new projects including their results. A large number of contributed papers has been strictly examined and selected by the editorial commitee to guarantee a high international technical standard.

The Second Automotive Simulation book contains the accepted papers which will be presented at the Symposium. The papers have been classified according to the following topics:

1. DRIVING SIMULATION and SIMULATORS
2. FEM for WHEELS AND TIRES
3. CRASHWORTHINESS
4. ENGINE DESIGN
5. DRIVE LINE and CONTROL SIMULATION
6. COMPUTATIONAL FLUID DYNAMICS (CFD)
7. SOFTWARE TOOLS
8. SUPERCOMPUTERS and HARDWARE TOOLS
9. CORROSION and ECOLOGY

Other papers covering

10. ACOUSTICS (NOISE)
11. ARTIFICIAL INTELLIGENCE

will be orally presented without a written manuscript in this book.

Authors from 8 countries will meet at the Symposium. They work for CARS and TRUCKS manufacturers, Computer Systems Industries, Universities and Research and Development Institutes, so that a broad spectrum of simulation know-how is covered: Theory and Applications, Hardware and Software, Research and Development.

We hope that this meeting will be scheduled traditional in Schliersee every October of the odd years to keep this stage for information exchange.

The editor is grateful to the authors for making possible the publication of this book, and especially to Mr. von Hagen and Ms. Raufelder of Springer-Verlag for the confidence in ASIMUTH`s ability and the excellent new simulation volume respectively.

My thanks also go to all of the ASIMUTH people who have been involved beyond everydays work in the promotion of the symposium. Without the contribution of all these people, the Conference could not have been materialized.

Munich, October 1991

Moshe R. Heller
ASIMUTH
Applied Simulation
Technology GmbH

Table of Contents

CRASHWORTHINESS

ENGINE

DRIVE LINE and CONTROL SIMULATION

COMPUTATIONAL FLUID DYNAMICS

SOFTWARE TOOLS

SUPERCOMPUTERS and HARDWARE TOOLS

CORROSION and ECOLOGY

Driving Simulations

Aspects and Perspectives of Contemporary Design Driver Methodology

M. Neculau, U. Kramer

Laboratory of Automation Engineering

FH Bielefeld, Zimmerstraße 15

D-4800 Bielfeld 1

Summary

Simulation tools are meanwhile well established and proven methods in design cycles of motor vehicles. The increasing complexity of car suspensions, drive line management systems, and other measures of improving car dynamics, as well as the growing importance of the customers' demands require an appropriate adaptation of the development tools.

Two main directions seem to be suitable to meet those challenges. Firstly, the experimental efforts may be enforced in order to obtain more detailled information about customers' demand profiles. Secondly, the human car driving performance is modelled and incorporated into computer programs for a closed-loop simulation of driver-car systems.

This contribution will be concerned with both approaches. It discusses some of the typical potentialities and tendencies of future car functions. The respective experimental test requirements, including the corresponding facilities, are derived. It emphasizes the methods of man-in-the-loop simulation and presents some criteria for its hard- and software realisation from an ergonomical point-of-view. On the other hand, results of the many years experience of both authors in the field of systems engineering driver modelling are surveyed and compared to those of other competing models.

Finally, a synthesis of both approaches outlined will be aimed at. This suggests a strategy connecting the man-in-the-loop simulation iteratively with the pure computer closed-loop simulation. It is expected that in this way the experimental efforts can be reduced to an inevitable amount, because experiments serve just as sample data sources of the computer closed-loop simulation.

Obviously, not all properties of motor vehicle dynamics can be covered by simulation techniques; therefore, some of the limitations of this methodology will be considered as well.

Introductory Remarks

The product specifications catalogues of passenger cars as they have been discussed in the automotive industry during the last years contain hundreds of items which are commonly structured in the following way:

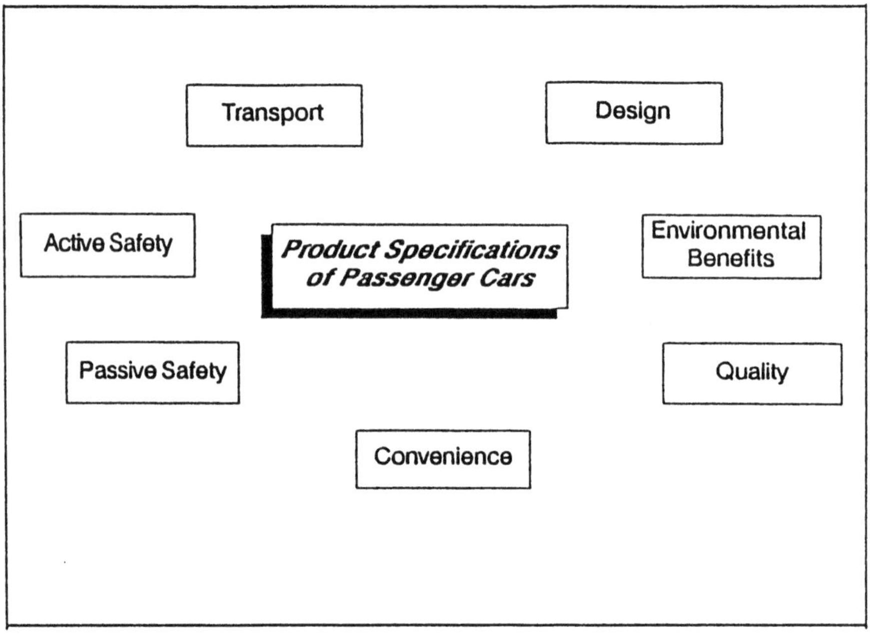

Fig. 1 Product specification criteria of passenger cars

It belongs to the engineers' main concern to transform these specifications into corresponding constructional features. Noteworthily, most of the specification criteria (about sixty per cent of them) cannot be captured just by numerical quantities, but either they may be expressed exclusively in terms of subjective evaluations and estimations, or they must be supported by them.

Going into more details, the percentages of "subjectivism" in the specifications criteria scheme would lead to the following figures:

- Transport: 15 p. c.
- Active Safety: 75 p. c.
- Passive Safety: 65 p. c.

- Convenience: 65 p. c.
- Quality: 60 p. c.
- Environmental Benefits: 40 p. c.
- Design: 85 p. c.

Since in all automotive companies the engineers are developing cars on the basis of quite similar means and methods, it can be supposed that those figures hold more or less for all of them.

Considering the main sources of knowledge about future properties of a car under development such as

- the feeling of experienced engineers and designers,
- results of precise calculations and simulations,
- tests of individual components in test stands,
- tests of individual components in experimental or conceptual vehicles,
- tests of prototype cars,

the question arises, if and to which extent the importance of subjective elements can be considered.

Design Driver Methodology

Without oversubtilizing the discussion about "subjectivism" in the car development, we have to be aware that

- subjective factors are of a paramount importance for the car development process,
- computer assisted design procedures have to be concerned with these subjective elements also,

- o design driver methods must bridge the gap between objective and subjective approaches, and

- o test rides (man-in-the-loop experiments) and simulations (closed-loop computer experments) must complement mutually.

Nonetheless, different face lines can be identified: the vehicle design strategy is oriented either to the *excellence of engineers* or to the *acceptance by customers*. Conflicts may occur not only between, but also within both attitudes, because desired and expected innovations are always in opposition to consolidating forces of preservation.

The design driver methodology rests upon the idea to provide a rationale

- o which may help to mediate between different points-of-view during the design phases (nothing else is meant by the notion of "subjectivism"),

- o which has to play the role of a communication platform for all the development partners, and

- o which is necessary to accumulate and to secure the know-how of the respective company.

Thus, three functional spheres can be distinguished:

- o moderator function,

- o distributor ("black board") function, and

- o accumulator function.

If we focus our attention to driving dynamics management, cockpit design, and in-traffic behaviour, it is evident that simulation methods can support the design driver rationale with the most important technical contributions.

Obviously, simulation methods are united with other computer assisted development tools by a common fate: they cannot simply substitute other research and development instruments, but they must complement and improve them. The generic advantages of such tools compared to proving ground or test stand experiments result from their variability, reproducibility, and cost effectiveness of evaluating the entire vehicle in early phases of development cycles. Their disadvantages are clearly reflected by the ranking where

- o simulations can never become better than measurements, and

- o measurements can never become better than "epicritical" sensations of test engineers.

However, the design driver rationale has also a characteristic feature that distinguishes it from other computer assisted methods: since the design driver approach includes both the man-in-the-loop experiments and the closed-loop computer experiments, it can "create" pieces of knowledge, whereas within in the simulation community the GIGO ("garbage-in, garbage-out") principle still holds.

Simulation Techniques

The vehicle engineers and designers are faced with a series of challenges (without reference to their ranking order):

- o shortening of development cycles,

- o control of increasing product complexity,

- o flexible responses to market demands,

- o necessity of an integrated system approach,

- o rapid prototyping of inventions and innovations, and

- o consequent customer-orientedness in all phases of car design and development.

Additionally, they have to master many technological imperatives resulting from discernible trends of future in-car systems. It can be expected that the following system properties will be of increasing importance:

- o *active* controls and actuators,

- o *co-operative and supportive* driver assistance,

- o *predicitive* driving environmental sensors,

- o *communicating* route and traffic information displays,
- o *safe and dependable* components and systems.

This list emphasizes that the modelling and simulation of the system driver-vehicle-environment will become of increasing importance too. Typical questions arising there might be:

- o Which are the "optimal" ranges of the drivers' performance?
- o How can the driving tasks be allocated appropriately?
- o How can we ensure a safe use of the reliable vehicle functions?
- o What has to be considered for an adequate design of driver-vehicle interactions?
- o How does the driver behave in critical situations?

From an analytical point-of-view, car driving can be seen as a sequence of goal-directed actions in achieving a given or desired final state (driving task). Each discernible intermediate stage between the initial and the final state is a driving state produced by the driver's manoeuvres. Variations of driving tasks may include:

- o goal-complexes (homing, route, direction, lane, speed, distances);
- o constraints (environment, wheather, road condition, traffic, illumination, vehicle characteristics, driver disposition);
- o requirements (situational and task complexity, time budget, discretion, object visibility).

Car driving is a complex of impressions and sensations which cannot be arbitrarily dissected into simple units (e.g. single-channel relations between stimuli and responses). This is the reason why since a couple of decades world-wide activities can be observed to develop and build driving simulators with increasing realism (the driver can, as to speak, "ride on the plotter" of the simulation computer). Yet the benefit of even highest-fidelity full-scale driving simulators must remain strictly constrained if the theoretical equipment is not developed accordingly.

As mentioned above, the design driver rationale is aimed at both approaches.

Driving Simulator

Three major areas of application have been identified: driving dynamics management, cockpit design, and in-traffic behaviour. The following list of applications is ordered with decreasing relevance:

- Application *required*: active systems (ABS, active suspensions), driving environment supervision (anti-collision systems, intelligent cruise control, distance warning);

- Application *advantageous*: communication systems (active beacons, fixbased route guidance systems, traffic information systems), novel cockpits (integrated displays and controls, voice i/o), accompanying research (accident causation, liability analyses, acceptance testing);

- Application *useful*: drive-line management (classification of drivers' behavioural patterns, control strategies), convenience (habituation and other long-term effects), safety (influences of fatigue, alcohol, drugs, etc).

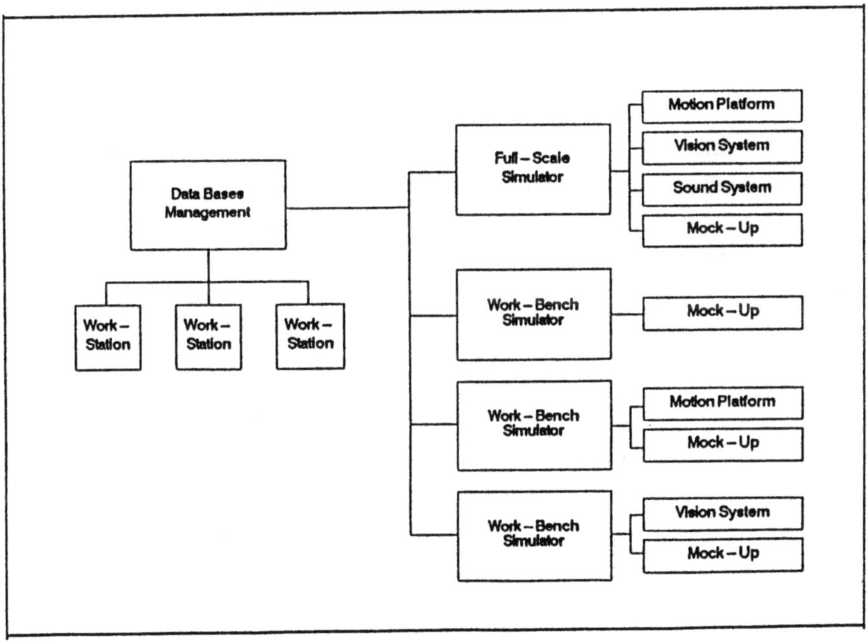

Fig. 2 Possible driving simulator configuration

With respect to the major applications mentioned, we recommend - a bit schematically - to proceed according to the following "shopping list":

- <u>Application</u>: **Cockpit Design**
  <u>Tasks</u>: analysis of any kind of deriver-vehicle interaction, especially secondary and tertiary driving tasks;
  <u>Simulator Configuration</u>: small-size simulator, mock-ups with precise imitation of displays and controls, "good-enough" graphics, no motion platform needed;
  <u>Expected Results</u>: differential diagnosis of user-groups (e.g. younger vs. elder drivers), direct comparison of alternative cockpit design concepts.

- <u>Application</u>: **In-Traffic Behaviour**
  <u>Tasks</u>: analysis of the multiperson-multimachine system "traffic" and of driving tasks to be embedded into reproducible traffic situations;
  <u>Simulator Configuration</u>: medium-size simulators, complete mock-ups with high-performance graphics, possibly multiple uniform simulator units, noise and vibration stimulation, motion platform stimulation advantageous;
  <u>Expected Results</u>: acquisition of representative exposition data from different user profiles, interaction of different road-users, traffic conflict and safety research.

- <u>Application</u>: **Driving Dynamics Management**
  <u>Tasks</u>: analysis and pre-optimisation of driving dynamics management system;
  <u>Simulator Configuration</u>: full-scale simulators, complete mock-ups with high-performance graphics, high-performance motion system, vibration and noise simulation;
  <u>Expected Results</u>: tendencies of dynamics quality according to particular standard driving and traffic situations and user demands.

The problem of a full-scale simulator might be seen in the fact that it ties up an enormous investment volume, although its practical use resembles typical single-user computer facilities, i.e. it is available for one application only at a given time. Therefore, it can be expected that future introduction strategies for driving simulators will have to take into account more strongly work-bench driving simulators covering specific development tasks. Possibly, configurations of driving simulation facilities will be installed as it is shown in Fig. 2.

Closed-Loop Computer Simulation

Fig. 3 shows the interaction between different information processing levels of a general design driver model.

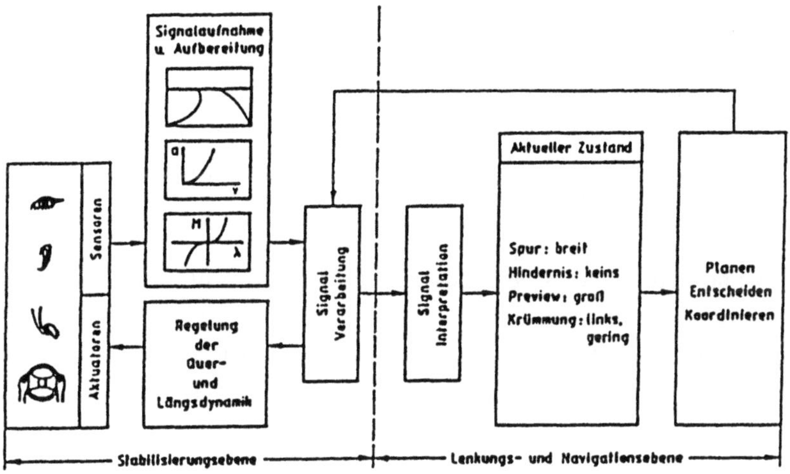

Fig. 3 Basic structure of a general design driver model

On the lowest information processing level ("Stabilisierungsebene" = stabilization level) the vehicle is controlled by highly automated manual processes which need precise information on the state variables of the process under control (i.e. describing variables of the vehicle and the driving environment as well). The information processing on this level consists mainly in the extraction and treatment of the relevant data from the surrounding in order to get the task variables needed.

The higher levels of information processing ("Lenkungs- und Navigationsebenen" = guidance and navigation levels) provide the lower level with reference variables which can be assumed to be taken from a knowledge basis. The corresponding decision-making and planning processes are inherently connected with a data reduction and, therefore, they are of an increasing uncertainty the higher the respective level is.

In order to build up a driver model which goes beyond the usual modelling of the stabilization level, the following aspects must be considered:

- o design of an appropriate data structure for economical storing mechanisms of large data quantities;

- o embedding of microscopic as well as macroscopic behavioural models into the data structure;

- o data reduction by means of a fuzzy partitioning of the relevant variables into classes of linguistic variables.

The methods of "artificial intelligence" are useful to manage large data quantities which are needed in the general design driver model. They allow the connection of declarative knowledge bases with procedural ones, e.g. by semantic networks. These semantic networks are composed of nodes (sometimes also named "concepts" or "frames") and edges representing the semantic relations between the nodes.

A concept may be taken for a form sheet which consists of two types of so-called slots: slots of the one type represent the relations between the concepts, whereas slots of the other type specify the concept attributes. Thus, a concept can be interpreted as a generic element which represents prototypically a class of objects, events, or other entities. The individual members of a class are named instances so that an instance is a filled in form sheet.

One approach of utilizing the knowledge stored in a semantic network is the construction of a control module which is exclusively oriented to the syntactical structure of the network. The basic structure of such a control algorithm is derived from an objective concept with its evaluation of attributes and structures.

At least five classes of concepts have to be distinguished:

- o driving task,

- o driving situation,

- o driving manoeuvre,

- o vehicle state, and

- o driver.

An example of the driving situation class concept is depicted in Fig. 4. The header ("KONZEPT") contains the name of the concept ("Fahrsituation" = driving situation). There are no generalizations ("GENERALISIERUNGEN") and no individualizations ("SPEZIALISIERUNGEN"). As an example of the instances ("INSTANZEN") the driving situation now ("Fahrsituation_Jetzt") is described, and two examples of attributes are listed ("Straßentyp und Ausbau" = road type and geometry, "Sichtverhältnisse" = visibility conditions). The body of the concept is closed with an end statement and the concept name ("ENDE Fahrsituation").

```
KONZEPT Fahrsituation
    GENERALISIERUNGEN -
    SPEZIALISIERUNGEN -
    INSTANZEN Fahrsituation_Jetzt
    ATTRIBUTE
    Straßentyp und Ausbau:
        TYP Konzept
        WERTEBEREICH Autobahn, Landstraße, Innerortsstraße
        ANZAHL-WERTE 3
        DEFAULTWERTE Landstraße
        BERECHNUNG Erkenne_Straßentyp & Ausbau
        AKTION-BEI-EINFÜGEN Berücksichtige_Straßentyp & Ausbau
    Sichtverhältnisse:
        TYP Konzept
        WERTEBEREICH sehr gut, gut, mittel, schlecht, sehr schlecht
        ANZAHL-WERTE 5
        DEFAULTWERTE gut
        BERECHNUNG Erkenne_Sichtverhältnisse
        AKTION-BEI-EINFÜGEN Berücksichtige_Sichtverhältnisse
        .
        .
        .
ENDE Fahrsituation
```

Fig.4 Example of a concept (class of driving situations)

Conclusions

Subjective factors of the car development and novel complex in-vehicle systems require improved methods of pre-evaluating the future properties of a car under development. The main tools of the design driver approach, man-in-the-loop experiments on driving simulators and closed-loop simulation experiments on digital computers, might bring the designers in a position to predict those future properties provided that the tools are refined correspondingly. This contribution outlines the direction in which the design driver rationale should be developed further in order to meet the requirements of the car development to be expected. The components of both tools are available, it will be, however, of decisive relevance, in which way these components are structured. We have given some proposals that seem to cope with this challenge.

References

Heller, M. R.; Kramer, U. (1990). *Driving Simulation - Task Requirements, Ergonomical Aspects, and Car Development Impacts.* International Symposium on Driving Simulation, Dearbon (MI), USA, Nov 8-9, 1990.

Kramer, U. (1985). *On the application of fuzzy sets to the analysis of the system driver-vehicle-environment.* Automatica, Vol. 21, No. 1, 101-107.

Kramer, U.; Marx, D.; Povel, R.; Zimdahl, W. (1987). *Technische Probleme und Lösungsansätze für das Forschungsprojekt PROMETHEUS der europäischen Automobilindustrie.* ATZ, 3/87.

Kramer, U.; Lerner, G. (1989). *Consideration of simulation tools for the development of vehicle complex systems.* 2nd European Cars/Trucks Simulation Symposium, Schliersee, Mai 1989.

Neculau, M. (1987). *Driver/Vehicle Modelling.* Forschungsbericht Nr. 247/87, Institut für Fahrzeugtechnik, Technische Universität Berlin, 1987.

The Driving Simulator: An Aid for Ergonomic Design of Car Interiors

IsabelleGUYARD

Renault, Research and Development, service 0076

Bernard Moteurs 2, 67 rue Galliéni, 92 500 Rueil Malmaison, France.

The driving simulator SCORE is designed to conduct human engineering studies on car interiors. It is a new testing mean to help ergonomists to assess the visual attention of drivers in connection with instrument scanning, and to measure any collateral driving disturbance. The visual system allows to render a sufficiently detailed road environment. The software allows to design any type of road scenery with the help of menus. Visual simulation is controlled by a mathematical model which describes the dynamic behavior of an automobile when the driver navigates through the visual database using traditional driving controls mounted inside a mock-up.

1. INTRODUCTION

The introduction of new types of instruments into cars has led to innovations in the data available to drivers concerning information on and monitoring of the vehicle as well as information about the road environment. However, these developments can result in an increased visual load with the attendant risk of distracting drivers from primary driving tasks [1].

Faced with this problem, Renault expressed the need for a design tool which would enable them to reproduce, for a representative population of drivers, a primary driving task (e.g. route-following or detection of external events) and to superimpose a secondary instrument-reading or use task upon it in order to quantify the degree of distraction from the primary task caused by the secondary one.

The desire to be able to reproduce experiments exactly from one subject to another without danger and before the prototype could be installed on a vehicle, led Renault to develop SCORE [Simulateur de COnduite automobile pour Recherche en Ergonomie (ergonomic research driving simulator)] in 1987. Designed as a high-performance measurement tool using visual reproduction, but not as costly as moving base global simulators, SCORE was designed in collaboration with INRETS (*).

(*) Institut National de Recherches sur les Transports et leur Sécurité
French Research Institute on transport and Related Safety Issues

2. BACKGROUND

As the task of driving primarily involves the visual sense, it was this aspect which was given priority when designing SCORE. The aim was to provide visual reproduction of road scenes in which drivers could operate in a realistic and interactive manner using conventional driving controls.

The accent was therefore placed on faithfulness of the reproduction in general. The specifications included:

- Reproduction of a route-following task on a road with varying geometry.
- Reproduction of detection of simple events connected with traffic:
  * following a vehicle with programmed indicators (brake lights, direction indicators).
  * Overtaking.
  * Oncoming vehicles.
  * Obeying three-color traffic and road signals.
- The necessity of a visual effort to accommodate the change from looking at the road to reading instruments.
- As faithful a reproduction as possible of the usable field of view in a light vehicle.
- Reproduction of image resolution compatible with that of the eye.
- Ensuring coherent interaction between input driving commands and movements of the vehicle within the scene.

This determined the design of the various elements of SCORE:

- a user-friendly data-base construction tool.
- Real-time, high-performance visual simulation.
- A dynamic, real-time model, reproducing conditions in a light vehicle, with sound.
- A realistic and modular vehicle model.

3. DESIGN OF THE ROAD DATA BASE TOOL

To provide display of existing or fictitious roads, or of projected roads, the technical solution for visual scenes is inclined towards three-dimensional synthesized image systems.

This approach gives realistic reproduction of a scene and any observer's eye position, useful for simulating different driver's heights (from driving a light vehicle to an HGV) or for giving an aerial view of the road for navigation use.

The data-base construction tool must be sufficiently "user-friendly" to allow a person who is not a computer specialist to be able to construct a road scene easily and quickly.

This objective was attained with the concept of the data base in the form of a linear "strip of road" including in succession in a transverse section the road, markings and near-side shoulder.

The characteristics of geometry and arrangement of these elements are chosen from a tree-structure menu.

Equally, it is possible to insert a file into the route which visualizes a junction of given complexity for which the user will specify the entry and out point(round-about, crossing...). Finally, when the ergonomist does not wish to reproduce a real route, it is possible to choose "random generation" of the route with an imposed minimum curve for bends. This type of solution allows for instant generating of several kilometers of road .

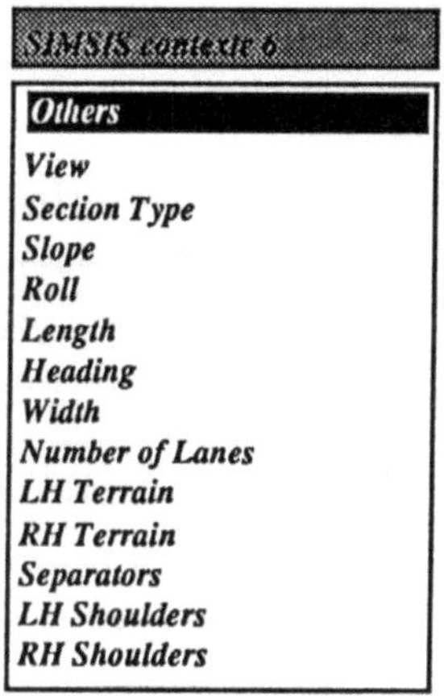

Fig. 1 Main Menu

Once this "strip of road" is created, a second phase consists in placing fixed objects in the scene (road signs, obstacles on the road, construction), in defining the conditions of visibility and colors along the route (good visibility, fog, day or night) and in specifying the scenarios of movements of the five mobile elements proposed for the scene. Each mobile element, pedestrian or vehicle, is specified according to its path of movement, variations in its speed, the laws governing lighting of brake lights, direction indicators or traffic lights.

Entering the path for a mobile element is done by driving with the mouse and entering the "passage points" on the keyboard. This makes it extremely easy to simulate lane changes, avoiding of an obstacle, negotiating of a bend, fork or intersection. The software then interpolates between the passage points. Entering other mobile element data is always done with the aid of menus and allows for construction of the following scenarios:

- the mobile element starts up at the same time as start up of the simulation or when the driver has reached a certain given distance, then continues its own evolution independently. This scenario makes it possible to impose strictly identical traffic conditions on different drivers.

- Variations in the speed of the mobile element are unknown to the driver. The only indication is approach or increasing of distance in relation to the vehicle. This allows for verification of whether a subject, when vision is removed from the road, retains visual awareness of the road environment and detects an approaching vehicle.

- The mobile element brakes for a given period and at a given rate of deceleration. Its brake lights are illuminated and its speed varies as a consequence.

This makes it possible to see whether a driver detects a mobile element braking in front when driving behind it. At present, triggering of braking depends only on a time setting, however, it could be triggered when the distance between the two vehicles drops below a threshold fixed by the controller of the experiment.

All of the scenarios have been defined with ergonomists with the aim of imposing a visual workload on drivers and estimating the level of visual awareness they retain of the immediate external environment (vehicle being followed, oncoming or overtaking vehicle) and of giving permanent indication of the capacity of drivers for acquisition of "priority" information while reading instruments.

To provide maximum flexibility, it is possible to associate different arrangements of objects and mobile-element scenarios on a given route. In particular, this allows ergonomists to take their subject through the experimental route without traffic or obstacles during the familiarization phase and then to add the experiment scenario to start the measurement phase.

Fig. 2 Exemple of road scene

4. REAL-TIME VISUAL SIMULATION

4.1. Interactivity

Once the scene has been constructed with the tool described above, it is possible to move around within it using the workstation mouse or driving controls.

In the first case, a sideways movement of the mouse on the pad simulates a change of heading for the vehicle and a longitudinal movement a change in speed.
This solution is immediately active and gives an instantaneous dynamic view of the scene.

The second case applies during the simulation phase.
This time, the heading and speed data are transmitted directly from the model vehicle to the workstation where they are interpreted for calculation of the image to be displayed.

4.2. Real-Time

In both last cases, a minimum image-refreshing frequency must be ensured to avoid drivers' visual sense being hindered. This basic constraint for ergonomic applications was taken into account by, on the one hand, limiting the complexity of the objects displayed and managing the different levels of detail according to their distance from the observer and, on the other hand, by systematic optimization in the software made possible by the architecture chosen for the database.

On most routes made by ergonomists, an image-refreshing frequency of 20 Hz is respected and is judged satisfactory. This rate remains closely linked to the type of machine used.
A more recent version of the graphics hardware provides a rate of 30 Hz for the same route.
In this case, the classic video standard is easily respected.

4.3. Resolution and Field of View

Another aspect of visual simulation vital for ergonomists and also depending on the hardware used, is the resolution of the image. Here a distinction must be made between calculated image resolution and displayed image resolution.
The calculated image resolution depends on the size of the machine's pixel-plan and of the cone of vision defined when the image is generated. This latter parameter is calculated so as to respect as well as possible the field of vision delimited by the windshield of a light vehicle and real-time performance; the larger the cone, the better will be the subject's visual impression. However, the greater the number of objects to be displayed the more the

refreshment rate will be degraded. The value of 60° in the horizontal plane, resulting from this necessity for compromise, gives, with the graphic station's 1280 pixels/line, a resolution of 3 minutes of an arc for the calculated image.

The resolution of the image displayed depends on its size. Use of a video projector provides an alternative to use of a monitor screen which requires more localized attention and limited movements of the eyes of an observer.

In order to give the subject a feeling of reality, the image must be large. SCORE, with its 3m50 width image and 1280 pixels/line resolution at the graphic station, gives a displayed-image resolution of 2.7mm.

Coherence with the calculated image resolution defined above, has led to positioning of the vehicle mock-up so that the observer's eye is at 3m20 from the projected image. This distance has been approved by the ergonomists whose aim is also to reproduce the effort of eyes adjustment required during driving when going from looking at onboard instruments which are close, to distant objects on the road.

5. ARCHITECTURE OF THE SCORE SIMULATOR

SCORE was designed as a mid-range simulator in order to produce an inexpensive tool for use by design departments. It was the visual simulation which required the largest investment, both for software and hardware. However, to attain a sufficient overall level of realism, it was necessary to develop, around the display system, other elements which consist of the host computer and sound reproduction system which complete the SCORE architecture.

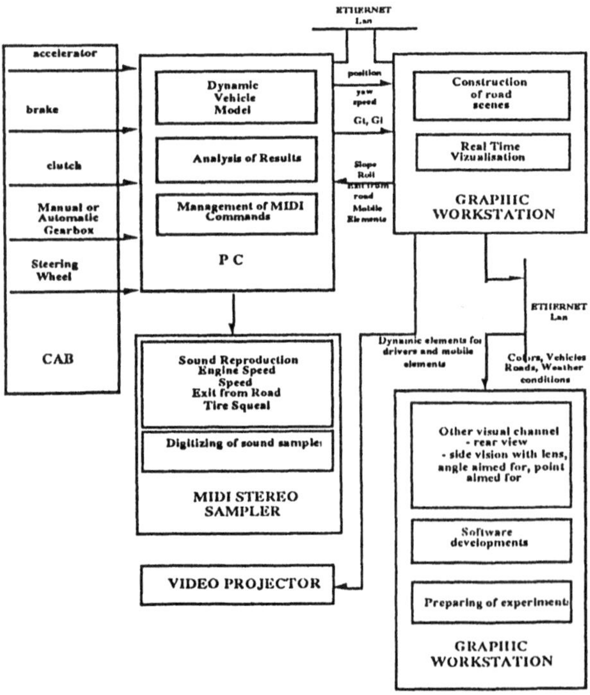

Fig. 3 SCORE Architecture

5.1. The Host Computer

The main simulator functions are managed by a central or "host" computer which sends the commands to the different dedicated systems for visual and sound reproduction. His tasks are as follows:

- Control of simulation loop between driving input commands and the different restitutions.
- Acquisition of the steering wheel angle, travel of brake, accelerator and clutch pedals and gear lever position.
- Animation of speed indicators and rev. counters.
- Dynamic vehicle calculations.
- Management of communications with graphics workstation.
- Management of interface with sound sampler.
- Storage of measurements.

Each of these points is therefore realized on a micro-computer, with standard boards and interface softwares.

5.2. Sound Reproduction

A sound system was carried out to restitute the sound surrounding inside the car. Based on the sampling technic of real car sounds, il allows to restitute engine noise, rolling noise, squealing of tires, noise of oncoming traffic and an alarm if the vehicle goes off the road. The necessary data are recorded in a car, using an acoustical head, and stored in a midi stereo digital sampler. The sampler is controlled during the simulation by the host computer, which sends to him the various parameters vai a MIDI interface.

These data are necessary for the driver to give a better impression of speed in the simulator and compensate for the absence of dynamic reproduction.

5.3 The Vehicle Dynamic Model

The visual simulation is controlled by a mathematical model discribing the vehicle dynamic. This model allows the pilot to evoluate in the visual data base, using the classical driving commands mounted in a half-car mock-up.

The choice of the mathematical model complexity was governed by 2 restricting factors :

- it must guarantee a sufficient handling quality in the experimental conditions
- it must run in real time, i.e in less time than the intergration time step.

To keep the model within reasonable complexity, and to keep the computation device within low cost, a 2 degree of freedom (2 ddl) model was set-up for the lateral dynamic. Chatelet and Pham [2] noted in their experiments that the vehicle driving frequency range was low, and that in these conditions, no noticable difference was observed between a 3 ddl and a 2 ddl model. As the roll is not a significant parameter for the pilot in normal driving conditions, the 2 ddl model is justified. This model receives as an input data, the computed vehicle speed from the longitudinal model. Both lateral and longitudinal models run on the host computer.

5.4. Measurement System

The SCORE simulator is a design study tool for ergonomists allowing for the creation of controllable and reproducible driving situations. SCORE has the additional advantage of facilitating access to a certain number of measurements which are more difficult to obtain in real driving conditions.

The measurements of relevance to ergonomists are those relating to route-following and detection of events in the external environment. The SCORE software allows for storage of the following parameters in a standard file:

- time,
- steering wheel angle,
- travel of brake and accelerator pedals,
- position of gear lever,
- vehicle speed,
- engine speed,
- various parameters linked to vehicle dynamics,
- the absolute coordinates of the vehicle in the scene,
- lateral deviation of route in relation to road center,
- road camber,
- leaving the road,
- a reference signal triggered by the experiment controller,
- the distance to mobile elements,
- the status of mobile element lights,
- speed of mobile elements.

These parameters are important for analysis of:

- activity of driver controls,
- driving performance,
- response time to different signals.

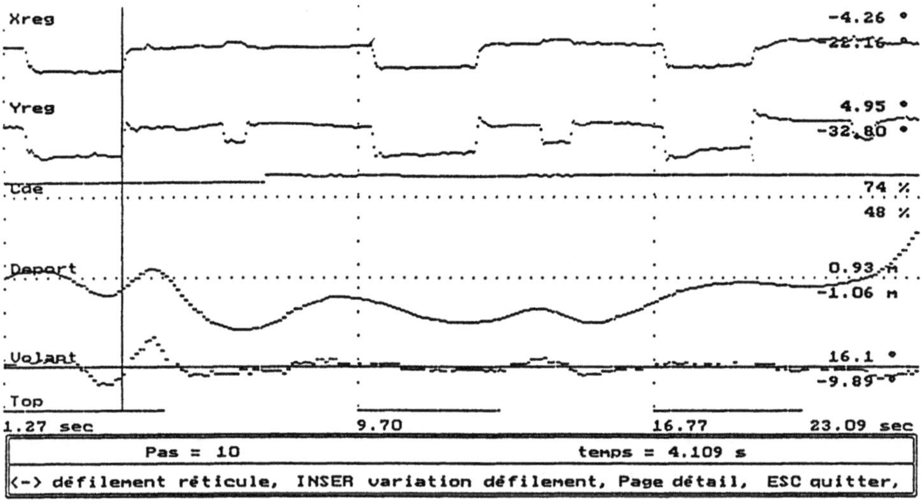

Xreg, Yreg: horizontal and vertical positions for driver's visual activity

Cde: Brake / Accelerator

Deport: Lateral deviation

Volant : Steering Wheel

Top : Reference signal

Fig 4 Graphic Representation of Visual Strategies and Driving performances

It is also possible to couple the SCORE simulator to different measurement systems such as eye movements detection, so that we can analyse in parallel the drivers visual activity and driving performance [3].

6. CALIBRATION OF THE TOOL

As we have already stated, the SCORE simulator is designed as a tool for the production of a visual workload for the driver. It is therefore logical to validate the system by quantifying the visual attention which it requires and by using results in conjunction with those obtained in real driving conditions [4].

Experiments are presently being conducted with this aim. Four positions have been defined for arrangements of instruments in the model vehicle [5]. A digital display device whose rate of display can be altered, is placed in each of the positions in succession.
The driver, proceeding at an imposed stabilized speed, is instructed to remain on the road, in the right hand lane.
When he feels ready, the driver is asked to read aloud a pre-determined number of figures on the display. Reading is to be done in a single operation without looking back at the road.
On the simulator, the deviation in route is measured directly ; in real driving conditions, it is mesured with the help of an optical system on a vehicle equipped with instruments.
The aim of comparison is to calibrate the driving task on the simulator (speed, width of road) to obtain performance levels comparable with those measured on the road.

Once calibration is completed, the first experiments on a prototype instrument arrangement can be carried out on the SCORE simulator, for analysis and transposition to real driving conditions.

CONCLUSION

The SCORE simulator is presently providing Renault ergonomists with an integral tool for the first phases in development of onboard instruments. It is a useful complement to the expert knowledge and other methods employed in ergonomics as it provides "objective" measurements on interface compatibility with driving and on driver's visual strategies. Its use makes it possible to forecast the results of tests in real driving conditions which will serve to conclude and confirm those made on SCORE.

Of course, it is clear that a simulator like this does not pretend to reproduce all of the complexities of the environment and of the acquisition of information observed on the road and does not, therefore, call into question the validity of final tests in real driving conditions.

Similarly, the absence of a moving base on SCORE and the low degree of complexity of the vehicle-dynamics model used, limit its use to the area of quasi-stabilized driving, or with only application of low dynamic workloads.

This point is not an important constraint for the applications intended for SCORE, as most of the ergonomic protocols consist on creating "standard" motor-way type driving conditions.

Finally, if the complexity of visual reproduction assumed to be necessary for automobile simulation in comparison with aeronautics has been, until now, a limiting factor on its development, recent upgrades in computer technology now allow it to be envisaged and to become accessible to an increasing number of users.

REFERENCES

1 Isabelle GUYARD - Hervé KERIHUEL - Evelyne SFEZ
Etude ergonomique de l'utilisation d'un écran tactile en conduite automobile
XXVe Congrès de la SELF - Octobre 1989

2 PHAM Anh Tuan - Alain CHATELET
Dynamique du système Conducteur Véhicule
Ingénieurs de l'Automobile - Décembre 1973

3 Hervé KERIHUEL
Présentation des outils de mesure des stratégies visuelles
XXVe Congrès de la SELF - Octobre 1989

4 Michel NEBOIT - Olivier LAYA
Etude de l'exploration visuelle du tableau de bord d'une automobile
Rapport final de contrat Renault / ONSER - Novembre 1983

5 F. HELLA - F. HARTEMANN
Etude de la durée de détournement du regard pour prélever des informations à l'intérieur du véhicule
Rapport interne Renault - Décembre 1982

A Load Vehicle Training Simulator

B. Bock
Krupp Atlas Elektronik GmbH,
W-2800 Bremen 44, Germany

Introduction

Real-time simulators for military and civilian airplanes have a long tradition as research and development tools in aerospace industry and for pilot training as being practiced by airline companies. Full mission flight simulators have attained lively interest in the media and are well-known to the public. Actually, these simulators play an essential role in professional pilot education, and their regular licence refresher trainings.

Some arguments supporting the advantage of using simulators for training purposes are:

- availability;
- cost efficiency;
- environmental concerns;
- no conflicts with normal traffic;
- new training capabilities;

To more or less extent equivalent arguments also hold for other training simulators, which are nowadays available as commercial products, e.g. tank trainers, ship handling or power plant simulators.

The advantages of simulator use are not so evident with truck or passenger car driving simulators, in particular when considering availability and cost of the simulator compared with the real vehicle.

Intended Application

Existing car driving simulators [3,4,5] and also recently proposed new driving simulator projects [2] have been designed for tasks like vehicle dynamics research, road construction optimization, car component development, or drivers' behaviour studies. Obviously, they are not intended to be used for training car drivers on a commercial basis.

Building a passenger car or truck simulator for driver training at reasonable cost so it can be afforded by driving schools is a new challenge. This is remarkably true since for this application the generation and display of realistic, complex traffic situations is essential, requiring a powerful visual system.

At Krupp Atlas Elektronik, we are going to build a truck simulator specifically designed for various applications in driver training courses. Simulator training shall become an attractive and cost effective part of initial and advanced driver training.

Actually, initial driver training will never be possible without traditional driving lessons in a real vehicle, because it is essential for the student to gain first, hands-on experience of the very complex real interactions between him, the vehicle, the road, and other road-users.

The great advantages of simulators exist in offering new training capabilities, not available before. In a simulator, the recognition and mastering of confusing, accident-prone traffic situations can be trained in a safe and efficient way. While in reality, it is generally impossible, or simply not practicable during the course of a normal driving lesson to provoke definite traffic situations or specific weather and road conditions.

To find a balance between cost and performance all subsystems of the simulator must be designed with the intended application in mind, i.e. according to the requirements of a road safety training facility. Whereas, for example, this training may also include unusual, rarely occuring vehicle conditions, the simulator will not be configured for tasks like evaluating the significance of subtle changes in vehicle design parameters, or testing vehicle dynamics at the physical performance limits.

The idea behind this concept is that normal drivers should better train to identify potentially dangerous traffic situations in advance and learn how to avoid them, than how to react in critical situations after they lost control of their vehicle. This applies in particular to truck drivers.

System Design Concept

In a simulator as well as for real driving, the driver needs sensory feedback in order to control the driving process. The required stimuli naturally occur for real driving in the forms of visual, vestibular, acoustic and tactile signals. In a simulator they have to be created by technical means to such an extent that the driver can manage his driving task and is not confused by apparent differences between simulated and natural cues.

The ultimate goal, to achieve perfect fidelity in simulation, has not been accomplished by even the most complex simulators. A realizable concept for a driving simulator first needs focusing on the most important sensory signals and then requires to compare gained system fidelity to extra cost for secondary perception cues.

Obviously, the eyes provide the essential sensory input to the human brain for driving. It is possible for a driver to control a vehicle reasonably well through visual images, only. Information about position, speed, and even some acceleration can be derived through the visual perception of the environment.

The simulator we envisage to build will not be equiped with a motion system to create vehicle acceleration cues. A full X-Y motion base or a tilted platform arrangement are ruled out by cost, maintenance, and safety reasons. However, to create some feel of road vibrations will be activated, and some degree of pitch and roll will be applicable to the cab to simulate spring deformation in its suspension. To add to the realism of the simulator, a real truck cab with fully installed, computer controlled instrument panel will be utilized.

In order to accomplish the degree of realism appropriate for the intended application, a high performace visual system will constitute the essential component of the driving simulator. High-fidelity acoustic cues and load control feedback form the steering wheel and the pedals will also contribute to the performance of the simulator.

Visual Subsystem

The visual system consists of the image generator system and the display system.

With existing technology two different ways of image generation are available, computer generated images and video (or film) based simulation. Computer generated images show the dynamic motion of the vehicle with high fidelity and provide a strong impression to the driver of his control over it. Prerecorded video based simulation is less expensive, but offers only limited capabilities to respond to the driver's control input.

For a prototype set-up of the proposed driving simulator, existing high performance computer image generators will be used. This system will serve to define the actual requirements for an appropriate image generator until final decisions will be made.

A proposed image display system covers a continuous 180 degree horizontal and ~30 degree vertical field of view; arrangements with front or rear projection can be imagined. In addition, two rear-view mirror displays are essential, since truck drivers heavily rely on these to assess the position of their own vehicle relative to the road and also to survey trailer movements and following traffic. Commercial video projectors will be used.

Audio Subsystem

High quality digital sound simulation will assist the driver to accept the simulated driving environment. There are basically two different types of sounds, continuous, repetitive sounds, and discrete sound events. The continuous, cyclical sounds that vary in frequency, harmonic content, and amplitude are relevant to indicating engine load, vehicle speed, wind, and different pavements. Discrete sound events mainly originate from passing traffic but also from the own vehicle. Both types of sound will be reproduced from sampled audio memory sequences.

Vehicle Dynamics Modeling

A modular software structure will be employed in modeling vehicle dynamics and subsystem characteristics. This approach will permit us to adapt the simulator to different needs and implement gradual improvements. The basic subsystems to be modeled are:

- drive train,
- brake system,
- steering wheel,
- tire-road characteristics,
- aerodynamic load.

At an elementary level, vehicle position, orientation, speed, and acceleration will be calculated with a linear model [1]. This model is known to give reasonable results on ordinary, dry roads under normal driving conditions with lateral accelerations below 0.4 g; a limit seldom reached by normal drivers, especially on trucks for which the rollover threshold may already impose a more stringent limit.

In modified versions of the dynamic model, non-linear tire-road characteristics and dynamic load transfers can be included empirically. Extra features specific for load vehicles, will require an extended vehicle dynamics model with more degrees of freedom. Some of the following issues are probably of interest for driver training:

- higher center of gravity of trucks, leading to a reduced rollover threshold;
- effects of different load conditions;
- off-tracking effects due to long wheelbase and multiple units (truck trailer combination, tractor-semitrailer);
- potential for jackknifing in hard braking maneuvers;
- retarded brake response;

Tire Model

The motion of wheeled vehicles is strongly dependent on the forces and moments applied to the vehicle from the tire-road interface. However, considering the planned purpose of this simulator, which is not research, we assume that a simple tire model will be sufficient.

A commonly used tire model is that from Dugoff et.al. [6]. For the case of braking and traction, the longitudinal forces are described by force-slip relationships ; slip being defined in terms of wheel spin rate, rolling radius, and vehicle velocity.

Tire side forces are given in terms of the tire cornering stiffnes. The model also yields a sufficiently correct description under conditions of combined longitudinal and lateral slip.

For the vertical tire loads, which enter the model as independent parameters, a static distribution may be assumed as a first approximation. Dynamic load transfers, induced by braking for example, can be included in a quasi-static manner as mentionend in [3].

In an upgrade for advanced training programmes, which may e.g. require the simulation of uneven road profiles, the tire model can be replaced by a more sophisticated one.

Conclusion

Design considerations for a load vehicle driving simulator to be used for initial driver and road safety training have been presented. In view of the rapid progress in computing and image generator technology, convincing, cost effective driving simulators suited for commercial use are likely to become feasible in the near future.

References

1. Mitschke, M.: Dynamik der Kraftfahrzeuge. Springer Verlag, Berlin, Heidelberg, 1988.

2. Haug, E.J.; et. al.: Feasibility Study and Conceptual Design of a National Advanced Driving Simulator. Iowa University, Iowa City, 1990.

3. Nordmark, S.: VTI Driving Simulator. Swedish Road and Traffic Research Institute, Rapport Nr. 267 A 1984, Linköping, 1984.

4. Hahn, S.; Kalb, E.: The Daimler-Benz Driving Simulator Set-up and Results of First Experiments. In Summer Computer Simulation Conference Proc., Simulation Councils Inc., San Diego, 1987.

5. Zimmermann, P.: Visual Simulation by Means of a Transputer Network for a Driving Simulator. Volkwagen AG, Wolfsburg, copyright Springer Verlag, 1989.

6. Dugoff, H.; Fancher, P.S.; Segel, L.: Tire Performance Characteristics Affecting Vehicle Response to Steering and Braking Control Inputs. Michigan University, Ann Arbor, Michigan, 1969.

Association of Automotive Characteristics and Functional Structure Concerning System Technique

C. Walther
Lehrstuhl für Raumfahrttechnik, TU München

Summary
The development of technical products changed very rapidly during the last years. An example for this is the automobile industry. Years ago the genius of a chief engineer kept the whole knowledge of the design process and the structure of the car in his mind. So he was able to decide everything very quickly and changes were well known in the whole company. Over the years the complexity of cars increased rapidly. Today, nobody is able to know everything about the whole system. These changes have caused a separation of responsibility for developing new subsystems to several managers. On the other hand the result of this decision increases the exigency for communication in information between all parts of a development department.

The product "automobile", its elements and subsystems have functions and characteristics. The producer of a vehicle develops functions, the user sees characteristic features. (Figure 1)

The central question of the research project "Association of Automotive Characteristics and Functional Structure concerning System Technique" could be formulated as follows:

How can functions and characteristic features of an automobile and even their influences on each other be described?
Which functions can be derived from required characteristic features?

This research project is an attempt to improve the development of automobile design by a better knowledge of the technical system description and the interactions between the departments of a large engineering team.

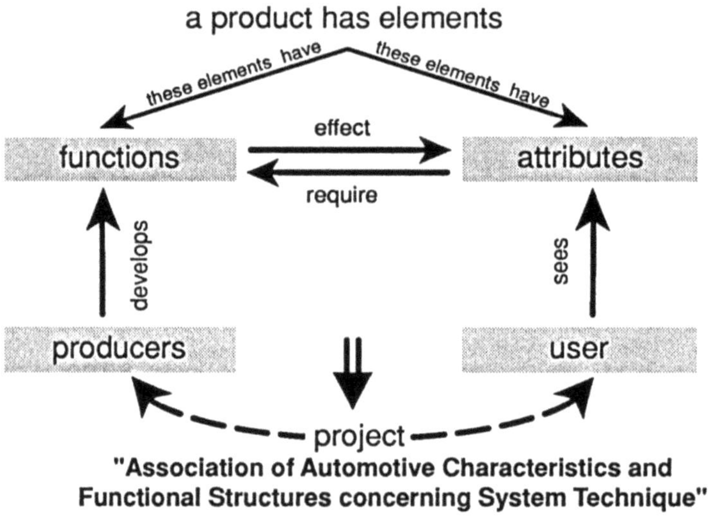

Figure 1:

Objectives of the Research Project

The major aim of any industrial producer should be the improvement and acceleration of developing technical systems and taking into account the requirements of market and customer. Problems arise because the development depends on several uncontrollable events and restrictions.

To reach these highly sophisticated aims several possibilities could be considered to improve developmental work in the automobile industry.

Reduction of developmental time

Under the pressure of market and competitors which decrease their developing time, producers, e. g. BMW or Daimler Benz with traditionally long term model cycles, are forced to reduce their developing times, if they do not want to lose part of their market-role.

Support for decision making

The management of development departments should be able to estimate the effects of their decisions, especially under the aspect of cost and time. Additionally, the effects on company organisation, workload of departments and last but not least the influence on system characteristics which are in the interest of customers, should be assessed much better.

Improvement of communication

The high linkage between the automobile elements requires an intensive communication between the department involved in the developmental work, to solve interface problems and decrease errors from the very beginning. For this objective a good description of the company organisation and their connection to the product designed are necessary.

Deference of engineers

The engineering department should be supported by software tools to decrease the workload imposed by standard tasks to be able to concentrate on their creative design missions. This requirement requires an easy-to-use software package which takes into account that all users can get all information they need to fulfill their tasks.

The major question now is: How can these aims be realized in a computer-based-tool to improve the development of a technical product?

To reach this objective the knowledge of the examined technical system should be improved in detail. So it is necessary to know everything about the parts of the complex dynamical system

buyer — market
product — technique
company — organisation

Technical requirements, relations between the system elements and structure of company organisations should be described in a way that everybody involved in the development team can get any information needed.

So for this project a complete description of the system "automobile" is required as fundamental principle for a cooperative process between all parts of development departments.

System Engineering Principles

A system has many different elements. These elements are designed by different engineers with varying education and experience (e.g. mechanical engineering, physics, informatics). This interdisciplinary work requires general definitions to get a common linguistic database, to avoid errors caused by misunderstanding. Anything in human life can be a system. Human beings form models, consisting

of elements, and the connections between them, to get a better understanding of a technical or social part of human life.
The following chapter describes basic terms of a general system definition.

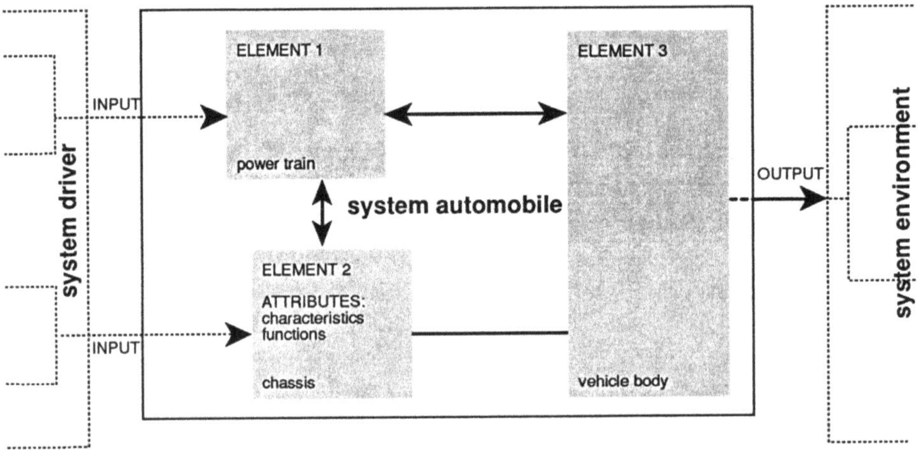

Figure 2: Definition of system element attributes

System

A system consists of a multitude of elements which have attributes and which are combined by relations. Thus a system should have a defined boundary which describes the parts belonging to the system or to its environment. The position of boundaries depends on the investigation and the designated results.

Element

The elements are part of a system (e.g. components, subsystems, a. s. o.). The structure of a system depends on the arrangement of elements. An element may be a system.

Attributes

The characteristic feature of an element or a system is the condition under a definite angle of view. Attributes of elements or systems could be characteristic features or functional tasks.

Attributes of state

These attributes describe the condititons of system or element attributes, e. g. geometry, mechanical or physical attributes. These attributes of state are defined and fixed in the engineering-phase and design-phase of a system or its subsystems.

Attributes of behaviour

These attributes describe the behaviour of a system or an element depending on technical profit, man-machine-interface and social influences.

Attributes of effect

These attributes describe the transfer of system inputs to system outputs, so they could be characterized as technical functions. Functions describe relations between two or more attributes of an element or different elements. These functions could be mathematical equations, enumerations of attributes belonging together or verbal information.

With these attribute categories a system could be described completely.

Relationsship

The relation is the directed or non-directed dependence between elements. These relations are flows of energy, material or information and dependences in space or time. With these technical dependences, relations inside a system and to the environment of system could be described.

The above definitions of the parts of a system are the formal criteria, which are the necessary basic items to get a complete description of any technical or other system.

However, problems are arising with regard to completeness and correctness of a system description. Therefore it is necessary to find methods to compensate the imperfections.

Principle Analysis

In the beginning of this research project a bicycle has been selected for principle analysis to investigate all oncoming tasks in development of a technical system. This example was especially used to define a uniform vocabulary and to recognize general problems. The reason why a bicycle was chosen is the analogy to the automobile, the actual field of investigation. The following information is needed during development of a system:

- identification of possible conflicts
- improvement of the transparency of system modifications
- harmonisation of characteristics of the whole system
- optimization of decision processes
- increase of communication between divisions

As a first step the element structure of a technical system, e. g. a bicycle, has to be defined to investigate changes of elements or their characteristic features. (Figure 3)

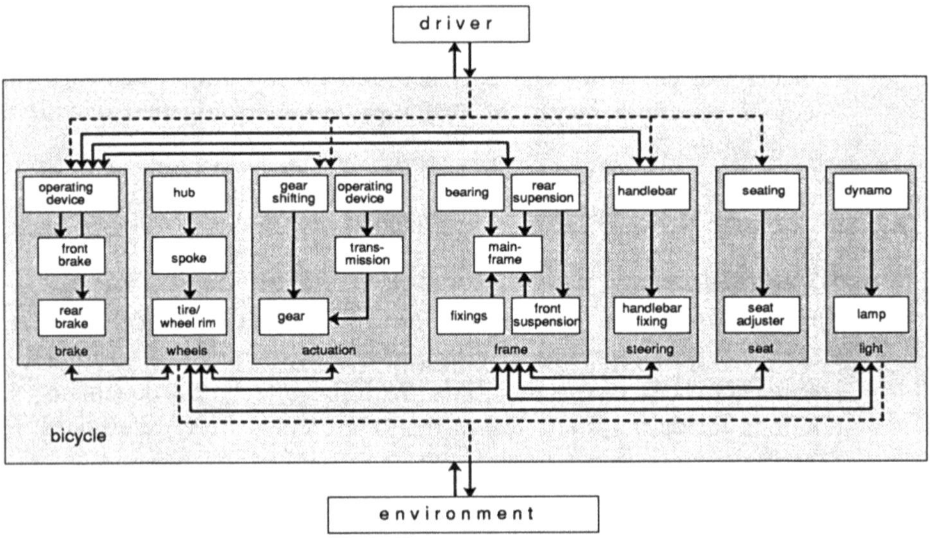

Figure 3: Structure of bicycle elements

To investigate the effects depending on changes in attributes or elements, a method was developed to recognize direct or indirect influence.

A question as for example the following one could be investigated with this method:

Which attributes or elements are direct or indirect influences by changing the comfort of sitting?

Figure 4 shows the complete method of the investigation of such questions. As results all interdependencies could be derived, especially between attributes and elements or elements and other elements.

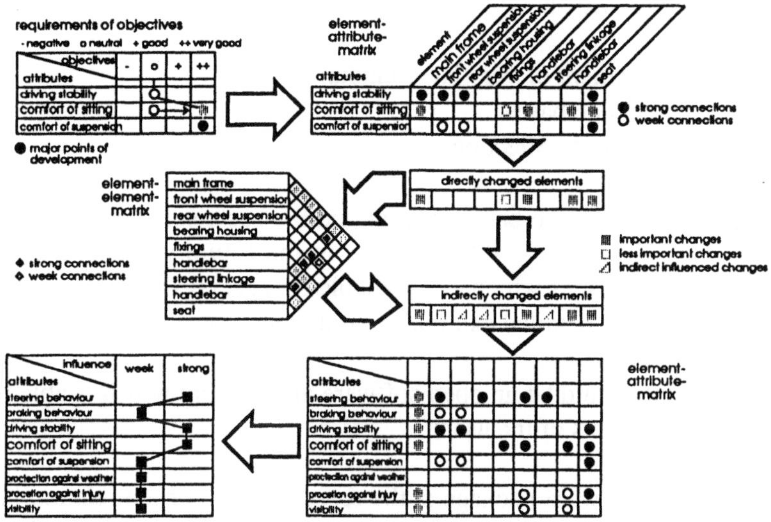

Figure 4: Influence of changes; example: comfort of sitting

Basic Concept

The previous investigations show that there are possibilities to derive methods which could be helpful for development of any system. These methods have to be translated into tools for the daily work. In case of complexity the only way to get an overview of the system automobile is to use computer-aided tools. Up to now many CA-Tools like CAD, CAT, CAE facilitate the engineering work. But an imperfection of most tools is that there is no common database behind them. A common database could possibly guarantee that people involved in a large project may get all information they need to do their work. In this way solutions should be worked out to optimize the common engineering activities.

Central point of this research project is to find computer-based tools to enable to describe all dependences of the system "automobile" in a common understandable way. The software tools should be used later on as a general database for information of management and engineers involved in development.

This method could be named Desktop Engineering (DTE), because every engineer in a company could get the information of interest on his desk via computer-aided tools, without problems caused by communication errors.

Surely, using this way of data exchange, there will arise new problems, as the following examples will show:

- data management of very large amount of data
- inconsistent databases
- expense for data collection

Therefore, the choice of the basic concept and the basic software used has to be done very carefully to avoid inflexibilities in the future work. It should have links to software packages which are available for special simulation tasks today or in the future, to use detailed external information if necessary.

The basic concept should be an open software architecture which is able to react flexibly on requirements of the users.

The kernel of the software package is the representation of the relation between elements and attributes of a system. To build up a simulation model of the relations of an automobile or any other technical system it is useful to know:

- elements and their relationships
- attributes of elements
- functions of elements depending on attributes

The task of system engineering is to find out a model based on generally usable description of a technical system and translate this model into a commonly understandable form.

Software for Desktop Engineering (DTE)

All system engineering methods should be combined in a Desktop Engineering software package which makes software-based support available for all parts of system design.

This proceeding should guarantee that all views of different people in an interdisciplinary development team could be coordinated in a better way, with information based on the same database. In addition it is necessary to design interfaces to standard or special software which are still in use. To get the best performance of desktop engineering tools it is decisive to have an easy-to-use user interface. Following task complexes should be supported:

phase-based tasks
- system analysis
- system synthesis
- system testing and variation
- system implementation

cross-section tasks
- documentation
- planning
- management

basic tasks
- text processing
- generation of graphics
- communication

To realize these requirements several software modules have to be installed:

Data management
The administration of several technical databases with general and special technical information of a project should be controlled by a data administration system, with functions to manipulate, combine and search through data. As a central element a data catalogue should be available which contains complete data definitions. The role of this database is to file all incoming data and make this information available to all users and all software applications.
To comply with all these requirements a relational or object-oriented database offers the greatest flexibility. Moreover, these databases should have interfaces for communication with other programs and run on different computer architectures.

Problem description
To define requirements for car development a formal definition language should be available which can translate a user's problem description into technical requirements. Even information of former design phases could be made available and this experience may simplify work for engineers considerably. This problem description language should have a user

interface with icon based programming possibilities and an intelligent history.

Simulation tools

For system synthesis, tools should be available which could enable engineers or even managers to transpose the requirements from system specifications and build up new simulation models. Now, available simulations tools cover special subjects and require intensive acquaintance. Since not only detailed information is necessary in design phases, new approaches in simulation modelling should be installed. This approach forces an easy to use simulation language, maybe even icon-based. The software should contain standard element descriptions, with all information needed for simulation. For further detailed investigations interfaces to special simulation products should be available.

Planning tools

For planning of development, methods are needed which collect automatically information of all differently designed system components and put this information in project overviews. Moreover, automatic procedures should recognize errors and work out recommendation of solutions. With this tool a continuous project observation and control is possible and, therefore, shorter reaction times depending on errors are imaginable. The planning teams could be integrated better in the whole development of a technical system, because communication errors depending on poor information could be excluded.

Management tools

The management needs information depending on the processes in the development of a technical system for decision making. For this task a computer-based tool could considerably improve the information flow in a company. Permanent access on the current state of development and forthcoming problems could cause a better decision basis. Decisions could be checked with simulation methods concerning their effects on the developed system without starting complicate proceeding all over the division of a developing team.

Vision

As result a desktop engineering software can arise which is able to describe the functional connections of an automobile and to support a development team by collecting and presenting useful information.

The software tool should be accessable to everybody in a development team so that everybody is able to get all information he/she needs to fulfill his/her task in an optimal way.

The highest level of a development tool could be the simulation of a complete vehicle from the very beginning at the concept phase up to the implementation and testing of subsystem.

> *A small example:* An engineer has an idea. With this idea he forms a new subsystem of an automobile named "XY 081". He changes the power train. The system informs him about restrictions, interfaces to other subsystems and tells him even the colleagues he should talk to to solve the problems arising. To be sure that the changes have little influence on other subystems, the engineer identifies the variations of parameters which have little influence on the whole system. These variations will be tested in a driving simulator and the results are evaluated with valuation methods on the desktop engineering tool. On this way results are generated without building up hardware in a short time.

Surely, there are some problems which will be difficult to solve. Such a large software package as the DESKTOP ENGINEERING SOFTWARE described above cannot be programmed in one step, but it is necessary to define all modules in the beginning of a new project and solve the interaction between them theoretically. This research project shows a new way of integrated engineering methods. Up to now some modules are realized to demonstrate and to investigate the effectiveness of desktop engineering methods.

But even if all dependences of a system were well known, such a tool would not be able to build up such a "noble piece of engineering art" like a bridge or a automobile by itself, because:

"System engineering is only the formalized part of engineering art. The non-formalized part belongs to the engineering art in the same way, and is equally essential or perhaps more essential: The creative human being."

(Prof. E. Igenbergs, TU Munich)

Literature:

/1/ IGENBERGS, E.:
Grundlagen der Systemtechnik: Skript zur Vorlesung
Lehrstuhl für Raumfahrttechnik, TU München 1990

/2/ EHRLENSPIEL, K.:
Umdruck2: Konstruktionstechnik I. Grundlagen und Methodenbaukasten zum funktionsgerechten Konstruieren
Lehrstuhl für Konstruktionstechnik, TU München 1989

/3/ BRILL, J.:
System Engineering Seminar June 1990
State of the Art, June 1990

/4/ Braess, H.-H.; STRICKER, R.; BALDAUF, H.:
Methodik und Anwendung eines parametrischen Fahrzeugauslegungsmodells
Automobil-Industrie (1985) 5, page 627-637

/5/ EHRLENSPIEL, K.; IGENBERGS, E.:
Bericht Vorstudie, Forschungsverbund Systemtechnik
Projekt A1: "Systemtechnische Zusammenhänge zwischen Automobileigenschaften und -funktionsstruktur"
Lehrstuhl für Konstruktionstechnik und Lehrstuhl für Raumfahrttechnik, TU München 1990

/6/ DAENZER, W.:
System Engineering, Leitfaden zur methodischen Durchführung umfangreicher Planungsvorhaben
Hanstein, Köln 1978/1979

/7/ WALTHER, C.:
Systemtechnische Zusammenhänge zwischen Automobileigenschaften und -funktionsstruktur, Fortschrittsbericht
Lehrstuhl für Raumfahrttechnik, TU München 1990

FEM for Wheels and Tires

Permanent Mold Aluminium Wheel Casting Optimization Via a Finite Element Simulation

P. D. Manhardt and A. J. Baker
Computational Mechanics Corporation (COMCO)
Knoxville, TN USA

Abstract

Permanent mold casting of aluminum wheels for the automotive market is a rapidly growing industrial sector. Foundry productivity is directly related to throughput, which can be dominated by the time required to cool the casting prior to mold opening. A finite element analysis prototype is developed and analyzed to parametrically assess active cooling options to decrease cast-to-removal time while maintaining quality.

INTRODUCTION

The automotive industry trend is towards use of die-cast aluminum wheels, for reasons of cosmetic appearance, safety and improved efficiency. The use of aluminum reduces the overall vehicle weight, resulting in efficient and superior performance. Statistically, 20% of all accidents are related to wheel/tire failures.

The permanent mold aluminum casting process occurs in four steps. The mo;ld surfaces are prepared and the mold is closed. The mold is then filled with liquid aluminum with pressure being maintained to fill regions of shrinkage, as solidification begins to occur. Following a predetermined length of time, the aluminum source is removed and further cooling and solidification take place. Finally, the mold is opened and the solid part is extracted.

Mold design directly affects two critically important aspects of the casting process. Unfilled shrinkage areas can occur if passages to melted regions solidify prematurely, thus blocking the flow of liquid aluminum. This can result in surface blemishes, or worse, internal weaknesses which, if undetected, can be potentially hazardous. The micro-structure of solidified aluminum is relatively unaffected by the cooling rate. With mold filling occurring on the *order of seconds*, while cooling time to removal is the *order of minutes*, a mold process which would hasten the solidification rate could have significant impact on production rates, hence costs, while maintaining high quality.

Unfortunately, aluminum wheels are geometrically complicated structures. The art of mold design for producing wheels therefore requires great skill and significant experience, making it a time consuming and costly process. Even so, an optimal mold cannot be guaranteed, and new designs are usually limited to *evolutionary*, rather than *revolutionary*, new concepts. Accurate computational simulation of the casting process can remove guesswork, and provides the tool for designing-in *passive* and *active* chill/refractory components, to hasten the solidification cycle while maintaining high quality. An accurate, efficient and versatile computational simulation model for support of mold design optimization is the subject of this paper

Specifically, a finite element discrete approximate solution algorithm is developed for the unsteady energy equation with full material properties variation including latent heats. The resultant prototype "ALUCAST" code employs an implicit time integration procedure, to accurately predict the unsteady temperature distribution within the casting and the mold elements. Solidification/remelt is included, as well as approximation to a mold-release agent as a nodal multiple degree-of-freedom. The benchmark simulation is for an axisymmetric geometry approximation of a generic rim-hub configuration, with both refractory and water cooled metal elements constituting the mold. The results of computational simulations are compared to available thermocouple data including parametric variations. The ALUCAST algorithm producing these data is now being implemented, as a *"template"* for the *AKCESS.*™ UNIX shell, for production analysis applications.

FINITE ELEMENT MODEL

The requirement is to obtain approximate solutions to the unsteady heat conduction equation with sources/sinks and variable/discretely-discontinuous material properties. The governing partial differential equation is

$$\rho c_p \frac{\partial T}{\partial t} - \nabla \cdot k \nabla T - s = 0 \qquad on\ \Omega \tag{1}$$

with Neumann and/or Dirichlet boundary conditions

$$k\nabla T \cdot n + q_n = 0 \qquad on\ \partial\Omega_1 \tag{2}$$

$$T(x_b) = T_b \qquad on\ \partial\Omega_2 \tag{3}$$

and an initial condition

$$T(\mathbf{x}, t{=}t_o) = T_o \qquad on\ \Omega \cup \partial\Omega \tag{4}$$

In (1), density (ρ), specific heat (c_p) and thermal conductivity (k) are distinct material properties, T is temperature and s is a source. In (2), q_n is the normal heat flux directed outwards from the appropriate domain boundary segment. For fluid convection, $q_n = h(T - T_r)$, where h is the film coefficient and T_r is the cooling water temperature.

A finite element computational (CFD) algorithm for (1)-(4) is based on a weak statement construction [1, Ch. 3]. Briefly, *any* approximation to the solution $T(\mathbf{x}, t)$ to (1)-(3) is

$$T(\mathbf{x}, t) \approx T^N(\mathbf{x}, t) \equiv \sum_{i=1}^{N} \Phi_i(\mathbf{x})\, Q_i(t) \tag{5}$$

and the *approximation error* resident in T^N can be extremized (minimized) by requiring that the integral of equation (1), written on T^N, be orthogonal to the (known) trial space function set Φ_i, $1 \le i \le N$, supporting T^N. The form of this "Galerkin" weak statement (WS) is

$$WS \equiv \int_\Omega \Phi_i \left(\rho c_p \frac{\partial T^N}{\partial t} - \nabla \cdot k \nabla T^N - s \right) d\tau \equiv 0 \tag{6}$$

which must hold for all i, $1 \le i \le N$.

The integrals in (6) are very difficult to evaluate in general geometries, and for spatially-variable material properties. Hence, a finite element (FE) approximation employs a discretization of Ω into a non-overlapping sum (union) of "finite elements" Ω_e, and selects compact-support (Lagrange) polynomials as the associated local form for the trial space $\Phi_i(\mathbf{x})$. Hence, (5) becomes

$$T^N(\mathbf{x}, t) \equiv T^h(\mathbf{x}, t) = \cup_e T_e(\mathbf{x}, t) \tag{7a}$$

$$T_e(\mathbf{x}, t) \equiv \sum_{i=1}^{E} N_i(\mathbf{x})\, Q_i^e(t) \tag{7b}$$

where superscript "h" denotes definition of the discretization Ω^h of Ω, "e" denotes data related to Ω_e, and $N_i(\mathbf{x})$ are the element set of basis polynomials.

The weak statement (6) then becomes

$$WS^h \equiv \sum_e WS_e = 0 \tag{8a}$$

$$WS_e \equiv \int_{\Omega_e} N_i \left(\rho c_p \frac{\partial T_e}{\partial t} - \nabla \cdot k \nabla T_e - s_e\right) d\tau$$

$$\equiv \int_{\Omega_e} \left[N_i \left(\rho c_p \frac{\partial T_e}{\partial t} - s_e\right) + \nabla N_i \cdot k \nabla T^h\right] d\tau + \int_{\partial\Omega_e \cap \partial\Omega_1} N_i q_n d\tau \tag{8b}$$

and the Green-Gauss divergence theorem has brought (2) directly into (8b). All integrals therein are easy to form, on the generic finite element domain Ω_e and its boundary $\partial\Omega_e$, and these expressions are summed (assembled) over all Ω_e producing (8a).

The unknown coefficients $Q_i^e(t)$ are now the time-dependent temperatures at the "nodes" of Ω^h, and the specific form of (8a) is the matrix ordinary differential equation (ODE) system

$$WS^h = [M]\frac{\partial\{Q\}}{\partial t} + \{R\} = \{0\} \tag{9}$$

In (9), [•] and {•} denote a square and column matrix respectively, and the weak statement residual $\{R\}$ contains all conduction, normal heat flux and source discrete approximate contributions. Equation (9) provides the derivative needed to evaluate a temporal Taylor series; selecting the θ-implicit one-step family for example, the terminal (computable) FE matrix statement is

$$[M + \theta\Delta t \frac{\partial\{R\}}{\partial\{Q\}}]\{\Delta Q\} = -\Delta t\{R\} \tag{9}$$

which yields the updated-time solution

$$\{Q(t + \Delta t)\} = \{Q(t)\} + \{\Delta Q\} \tag{11}$$

containing the temperature (approximation) at all nodes of Ω^h.

The ALUCAST prototype is coded for the problem statement (1)-(3) expressed in axisymmetric coordinates. The finite element domain shape is a quadrilateral containing four vertex nodes with associated nodal temperatures. The density and conductivity are assumed discretely discontinuous by regions of Ω^h, and $c_p = c_p(T)$ is defined to handle the latent heat effects during solidification/remelt. The matrix statement (10) is iteratively solved using a grid-sweeping approximate factorization linear algebra procedure [1, Ch. 8].

DISCUSSION AND RESULTS

The FE simulation algorithm has been validated for the prototypical aluminum diecast wheel sketched in Fig, 1. The axisymmetric assumption limits the hub cross section be averaged. Hence, the wheel geometry that is "seen" by the FE analysis is the cross-hatched region in Fig. 1.

The first step to a simulation is to define a suitable discretization for the wheel cross-section, and the surrounding mold. The associated algebraic procedure utilizes geometric blocks, called "macro elements," to segment the generic regions, i.e., aluminum, steel, iron, refractory, coolant (water), air, etc., of Ω. Figure 2 illustrates the developed macro element discretization Ω^H, with regions appropriately labeled. Each Ω^H is assigned a materials property table, and is then finely discretized to form a portion of Ω^h. (This is a simulation definition function that is highly appropriate to become CAD-based, as the development concept moves to production use.)

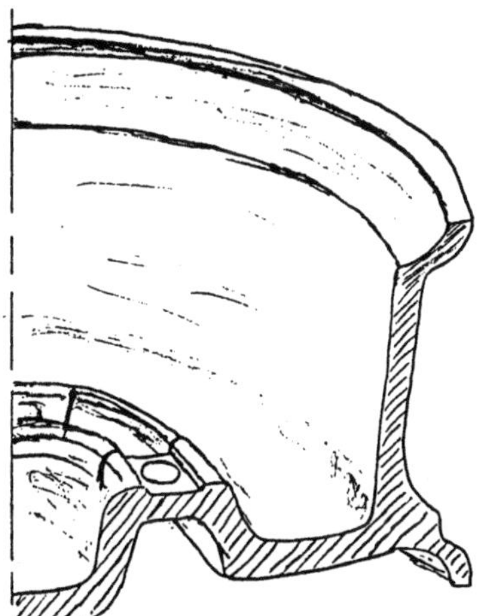

Fig. 1 Axisymmetric casting model.

The selected developed discretization defined a 65×65 nodal mesh which is geometrically quite contorted. A rastor graphics print screen output procedure was developed for rapid viewing of data in the transformed (nodal) space. Using a rounded hexadecimal scale, the complete mesh nodal temperature data field could be screen displayed in ±50° intervals for the temperature range $1300 \leq Q_i \leq 100$ °F. Figure 3 shows the rastor initial condition temperature distribution, with macro boundaries added to help assimilate the presentation format. The central maximum temperature for character "D" is 1317 °F, the

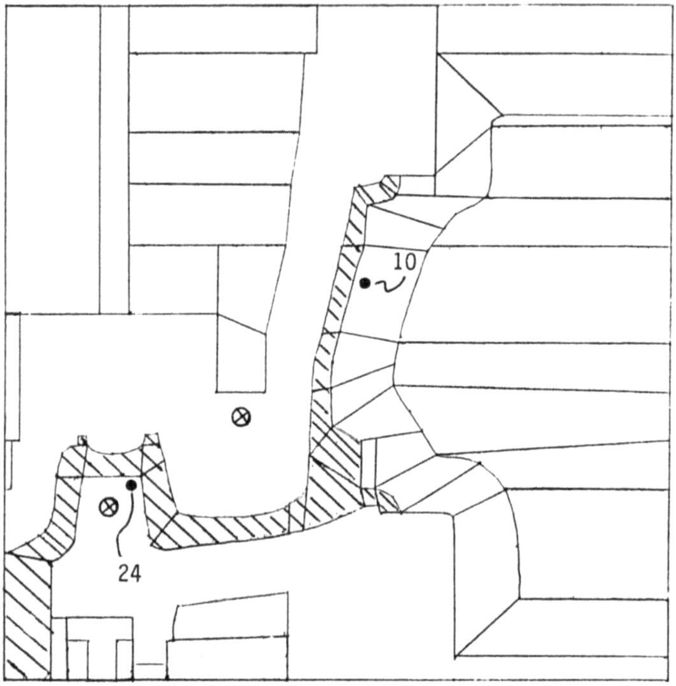

Fig. 2 Macro element discretization of wheel and surrounding mold, ⊗ denotes water channel, ● locates a thermocouple, wheel casting is cross-hatched.

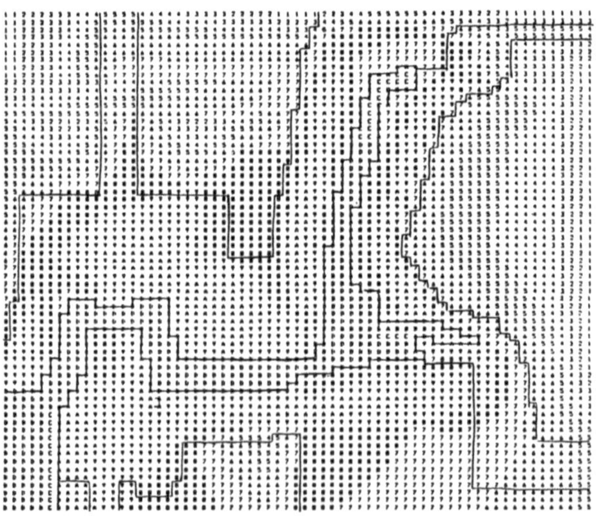

Fig. 3 Simulation initial conditions, hexadecimal printfield, T_{max} = 1317°F.

initial aluminum melt temperature. The coolest part of the mold (character "1") is initially at water jacket temperature (100°F).

Figure 4 summarizes the FE simulation results as rastor snapshots in time, for the case of the external water jacket not operating. After Δt=70s from mold fill, melt solidification locations are signalled by first appearance of asterisks in the print field, Fig. 4a. The freeze temperature is 1040 °F, and "9-A" is the indicated neighborhood melt temperature range. The sprue/riser temperature remains at fill condition, "D-C", and the wheel web and lower rim are largely at "C". The average bulk temperature of the entire wheel-mold assembly is output as 833°F.

Solidification is exothermic, and the resultant localized heat addition is predicted to cause selective remelting in the simulation. This cycling appears completed by Δt=148s, Fig. 4b, wherein the solidification process in the wheel and wheel-hub regions is quite advanced. The sprue/riser temperature "C" is down to 1210°F. Figures 4c-4d show the temperature fields at Δt=212s and 286s; the final sprue/riser temperature "C" is 1178°F, the rim solidification is almost complete, and the lug bolt face is beginning to freeze. The bulk temperature has decreased to 820°F.

Select foundry experimental temperature data were available, for this wheel geometry, for comparison to simulation prediction with no external water jacket. These laboratory data were acquired with a mold release material coating surface applied before fill. Comparison between experiment and CFD simulation was not adequate until the high thermal resistance character of this coating was included. The needed properties data were assimilated from Fig. 5, whereby the mold release thermal resistance could be estimated from coating thickness. Including this into the FE model was accomplished using the multiple nodal degree-of-freedom technique, since the actual coating thickness (microns) was not resolvable within the mesh Ω^h.

Figure 6 documents the resultant favorable temperature history comparison obtained for the CFD simulation with no external water jacket. The locations of thermocouples 10 and 24, and the internal water cooling pipes, are noted in Fig. 2. Overall quality agreement is indicated, except in the first minute or so of the simulation, when the CFD transient appears sharper than the data. This early time portion of any simulation is the most sensitive to initial conditions, recall (4), which the mathematics model assumes are known data. In truth, these data are ordinarily only approximately available, as was the current case, where the CFD initial condition was generated via interpolation of

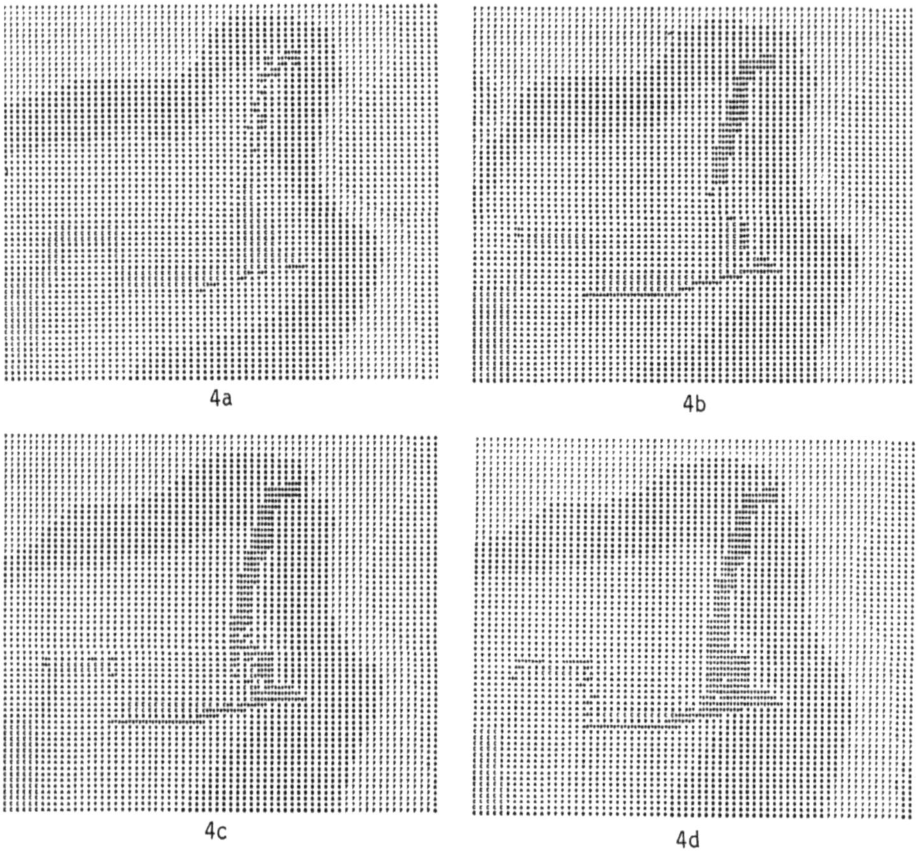

Fig. 4 Hexadecimal printfields of ALUCAST simulation temperature fields.

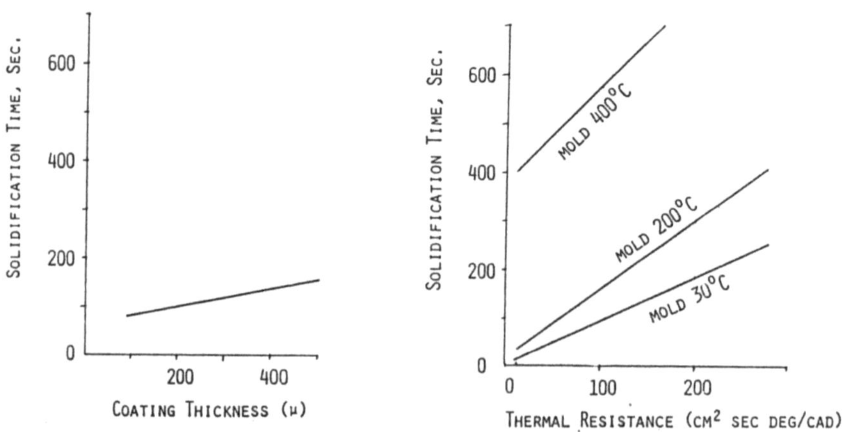

Fig. 5 Mold release materials properties characterization, from Y. Ohtuska, et al.

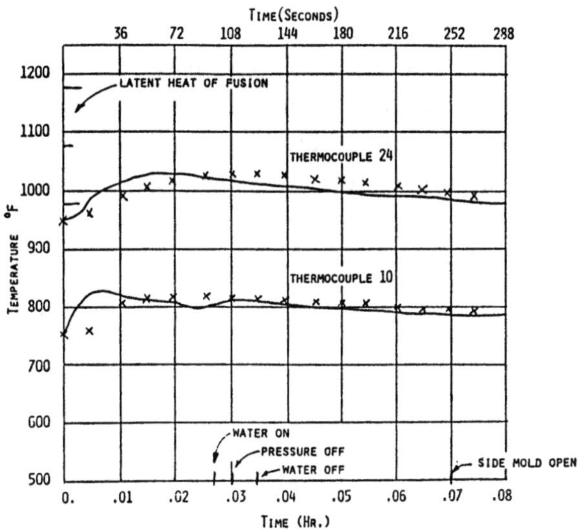

Fig. 6 Simulation comparison with experimental data, I.C. is 600°F air.

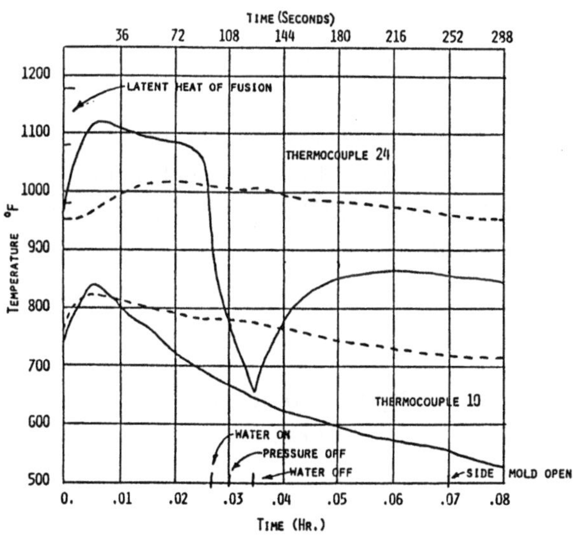

Fig. 7 Simulation results variation with external water jacket operation, I.C. is 100°F air.

relatively few experimental data points. Parametric studies on initial conditions could certainly be conducted to assess model sensitivity.

Once benchmarked, larger scale parametric variations can be conducted on the CFD simulation model, to assess prototype optimization protocols. For example, Fig. 7 graphs the temperature histories at nodes located at thermocouples 10 and 24 for the previous case (dashed lines), and for the simulation of an external water jacket. The early transient at node 10 is relatively unchanged for $\Delta t \leq 200s$, but thereafter the mold temperature steadily decreases at a much faster rate. Interestingly, the overall larger radial mold heat flux initially increases the temperature at node 24, via heat conducted from the riser. This temperature is then sharply decreased, when the internal cooling water is turned on, and then sharply increases when this heat sink is shut off. It then levels off at a temperature level approximately 150°F lower than the non-jacket case.

SUMMARY AND CONLUSIONS

A finite element simulation algorithm has been developed to support automotive permanent mold operation protocol optimization. The critical factors in its design include multiple media materials capabilities, solidification latent heat effects, and modeling of a mold release agent. For an axisymmetric geometry model, the ALUCAST code simulation results compare quite favorably with available experimental data. Parametric variations on prototype cooling protocols can be easily designed and executed, in the quest for an optimum casting operational procedure. The finite element algorithm is now being developed as a template" for the *AKCESS.*™ UNIX production shell for engineering graphics workstations.

Reference

1. A. J. Baker and D. W. Pepper, *Finite Elements* 1-2-3, McGraw-Hill Publishing Corp., New York, NY 1991.

Finite Element Analysis of Rubber Components in Hutchinson

D. BENOUALID
HUTCHINSON S.A.
Centre de Recherches - BP 31 - Chalette sur Loing - 45120 - FRANCE

I. WANDER
APEX TECHNOLOGIES
43 rue Froideveaux - 75014 - PARIS - FRANCE

INTRODUCTION

Founded in France in 1853, Hutchinson, in which the French petroleum Group TOTAL, has a majority shareholding, is the European leader in the field of industrial rubber, in all its forms. Its activities are built around three major markets :

- automotive, (Hoses and tubings, profiles, antivibration components, transmission systems, precision molding...),
- industrial (sealants, defense and safety systems, antivibration components...),
- consumer products (gloves, baby and healthcare, bicycle tires...).

These activities concern five core businesses in which Hutchinson possesses acknowledged expertises :

- extrusion,
- mouling,
- rubber-to-metal adhesion,
- latex dipping,
- boot manufacturing.

Hutchinson has production facilities in a number of countries (19 plants in France and 15 outside the country) and sells worldwide. Its products are both dependable and innovative, exclusive and universally known under such brand names as Aigle, Fit, Hutchinson, Lif, Le Joint français, Mapa, Nuk, Baby Relax, Paulstra, Spontex... They are born of the know - how of its manufacturers and the skills of its research teams. At the heart of this achievement lies the Research Centre at Montargis near the Group's earliest established plant. The Scientific Calculation Service which is one of the Research Centre Services, is in charge of évaluating the mechanical behavior of rubber components (such as rigidity, stress and strain concentration...) by numerical simulation using finite element method. Structural analysis are currently performed on rubber components to permit a better understanding and the design optimisation.

Numerical modelling of rubber components is a complex task because of the nonlinearity of material response, large deformation, etc. To succed sofisticated software and powerfull hardware is required together with an accumulation of experience. Because of the importance of the calculations on the delays in the developpment of newports and their quality, the R/D policy is to expand continously the facilities required. Recently a CONVEX C 240 (200 Mflops) supercomputer was added to the existing network of Silicon Graphics workstations and servers. The software developpment team is constantly enhancing the capabilities of an inhouse code which is used for rubber computations. In addition commercial codes are used for graphic pre/post processing and calculations which does not involve rubber mechanics such as extrusion, fluid dynamics and acoustics.

MECHANICAL ASPECTS

Rubber is a polymer which can undergo large elastic straining without residual deformation (elastomer). Rubber consists of long molecular chains which are linked together by relatively weak chemical bonds. The bonds are created during a vulcanization process. Different additives (fillers) are usually added to enchance particular mechanical properties (such as temperature resistance for example) ; from these additives the carbon black is the most known.

Initially the molecule chains have random orientation and hence the material properties are isotropic. But during straining, the elongation of the chains can create anisotropic material behavior. Hysteresis (dissipation due to internal friction) is a very strong function of the fillers present in the rubber compound, the temperature, and the strain level. Creep strain rate can be developped under constant stress conditions or stress relaxation may occur under constant strain conditions. At low temperature, this phenomena is dominated by mechanics of the chain deformations, at higher temperature the creep is controlled by chemically induced changes in material properties. Creep produces residual deformation, but the elastic properties of the rubber could also be modified by either aging or mechanical damage which brake part of the chemical bonds between the molecular chains.

The characteristic of the rubber components which is mostly used is their high elasticity. It permits to store (and recover) mechanical energy on a large deformation range. Combined with the rubber incompressibility, the high elasticity makes it possible to design springs with different rigidities in different directions of force application. The damping properties of rubber are used to dissipate energy in the vibration isolation devices, but damping can become a negative factor, when excessive heat build up is generated due to poor rubber conductivity.

Variety of industrial applications creates many different loading conditions and failure modes of rubber components. Unlike in linear behavior of metals, no unique failure criteria can be established.

In some cases the usual stress or strain criteria are reasonable to apply, but even then, much of experience is required since it is very difficult to develop a databank of material properties due to multitude of compounds. It is a constant challenge to the experimentalist to provide consistent results inspite of the problems involved. Continuous enhancement of test specimens and procedures is required to get stable and repetitive results. Currently in Hutchinson several testing procedures are available (related to the mechanical behavior) : measurements of elastic properties, creep, fatigue and crack propagation.

NUMERICAL ASPECTS

During a finite element analysis, two basics quantities are calculated :

- the internal force vector $\int_V B^T \sigma \, dv$

  (required in both explicit and implicit time integration)

- the tangent stiffness matrix $\int_V B^T DB \, dv$

  (only in implicit time integration)

Here B denotes a discrete gradient operator (shape functions derivatives) which is a consequence of a numerical approximation and discretization procedure. The material behavior is described by relations between the stress tensor ( ) and the strain tensor. These relations are usually called constitutive (or stress-strain) relations. The stress and the strain are related by means of a constitutive matrix D. This matrix is constant for linear material behavior. For nonlinear elastic material (for which a stress free reference state is defined), D is a function of current deformation state only. For inelastic processes it can be a function of additional history variables.

Rubber is usually considered as hyperelastic material, i.e. existence of a strain energy density function is postulated, from which the stress-strain relations (expressions for stress and constitutive matrix) are derived. The energy density function has to be conform to several requirements to be utilizable in the calculation. First, the derived stress-strain relations have to fit the experimental data in a sufficient range of strain and for different loading types (tension, compression, shear...). If this requirement is not fulfilled, no reasonable simulation of real behavior could be performed. The second requirement is that the constitutive matrix has to be positive definite on the complete range of application. Otherwise, numerical instabilities could occur during the analysis. This condition is equivalent to the statement that the stored energy should be positive for any possible strain. The third requirement is the objectivity of the stress rate. Objectivity is one of the fundamental principles in mechanics. The stored energy density function in nonlinear elasticity is a function of deformation gradient F ; i.e. $W = W(x,F)$.

It is said to be objective (frame invariant) if W (x,F) = W (x,QF) where Q is any proper orthogonal transformation. It can be shown that this condition will hold if W is a function of C = F F (does not change under superposed rigid body motion). The rate of Cauchy (true) stress is not objective, but the second Piola Kirchoff is. In hyperelastic calculations usually the second Piola Kirchoff stress is derived, and then the Cauchy stress is obtained by means of transformation using the deformation gradient F.

The most studied form of the strain energy density function is the polynomial expansion in the deformation gradient invariants. From these, the most widely used (and known) the Neo-Hookean and Mooney-Rivlin forms which use respectively one and two terms. Baltz-Ko form of strain energy density is said to be good to simulate the behavior of porous rubbers (sponges).

Following are some examples :

. **Neo Hookean :**

$$W = \tfrac{1}{2}\,\mu\,(I_1 - 3) \qquad \mu = \text{Shear modulus}$$

. **Mooney Rivlin :**

$$W = C_{10}\,(I_1 - 3) + C_{01}\,(I_2 - 3)$$

. **General polynomial form :**

$$W = \sum_{i+j=1}^{n} C_{ij}\,(I_1 - 3)^i\,(I_2 - 3)^j$$

. **Ogden :**

$$W = \sum_{n=1}^{3} \frac{\mu_n}{\alpha_n}\,({}_1 + {}_2 + {}_3 - 3)$$

In addition to the distorsional energy, the volumetric part has to be considered. Rubber is a quasi incompressible material, i.e. the bulk modulus is much higher than the shear modulus. Such a case cannot be treated numerically with standard displacement finite element approach, since severe mesh locking will occur. The two methods which are mostly used today are the mixed method in which the pressure is discretized as an additional variable, and the penalty method in which an incompressibility constraint is imposed by a function which "penalize" the deviation from the required constraint. We use the penalty method with selective reduced integration on the penalty term. In our experience the penalty method is more practical also for other types of constraints, such as contact surfaces (constraint of non-penetration).

The finite element discretization results in a system of equations which is solved by Newton Raphson iterative procedure. In our experience for most of cases involving high strain levels, the full Newton (reforming and cholesky factorization of the stiffness matrix each iteration) is the most efficient method.

The large deformation of the parts can create severe mesh distortions. In our experience for large scale computations the simplest elements which use the linear stage functions are the most appopriate to use. The are more robust in the sense that the global and local solution convergence is better and many non physical effects resulting from bad numerics are prevented.

EXAMPLES

Figure A is an example of an engine mount composed o rubber envelopping three composite plates. We can see the deformed shape and the high strains that exist below the lower plate. the force/displacement curve of the part is also shown. This analysis made it possible to increase significally the service life of the mount by optimizing different geometric parameters.

Figure B demonstrates three configurations of a seal. Once one half of the seal is modeled because of symetry. The seal undergoes compressive stress state arising from the movement of the bottom plane. These analysis were performed in order to validate the behavior of the seal with different tolerances, using different friction coefficients. From the rigidity curves depicted in figure B.4, one can see the rapid change in the behavior of the rubber when the volume is filled.

Figure C shows several results from the analysis of a belt. Fitting of the belt on the pulley create bending and tension forces. In addition the teeth undergoes local bending due to the torque transmission.The belt is reinforced by different types of composite or metallic cords. The rigidity for two configurations is presented in figure C.4.

On the figure D, the mesh and the results from the analysis of an engine mount are presented. This is a hydromount and therefor the geometry of the mount is complicated by the presence of the channels. The objective of this analysis was to calculate stresses and strains under severe loading conditions including contact between different parts. This model consists of more than 10 000 modes and the final state was archived by 60 load increments. The rigidity curve shows a very good agreement with the test data.

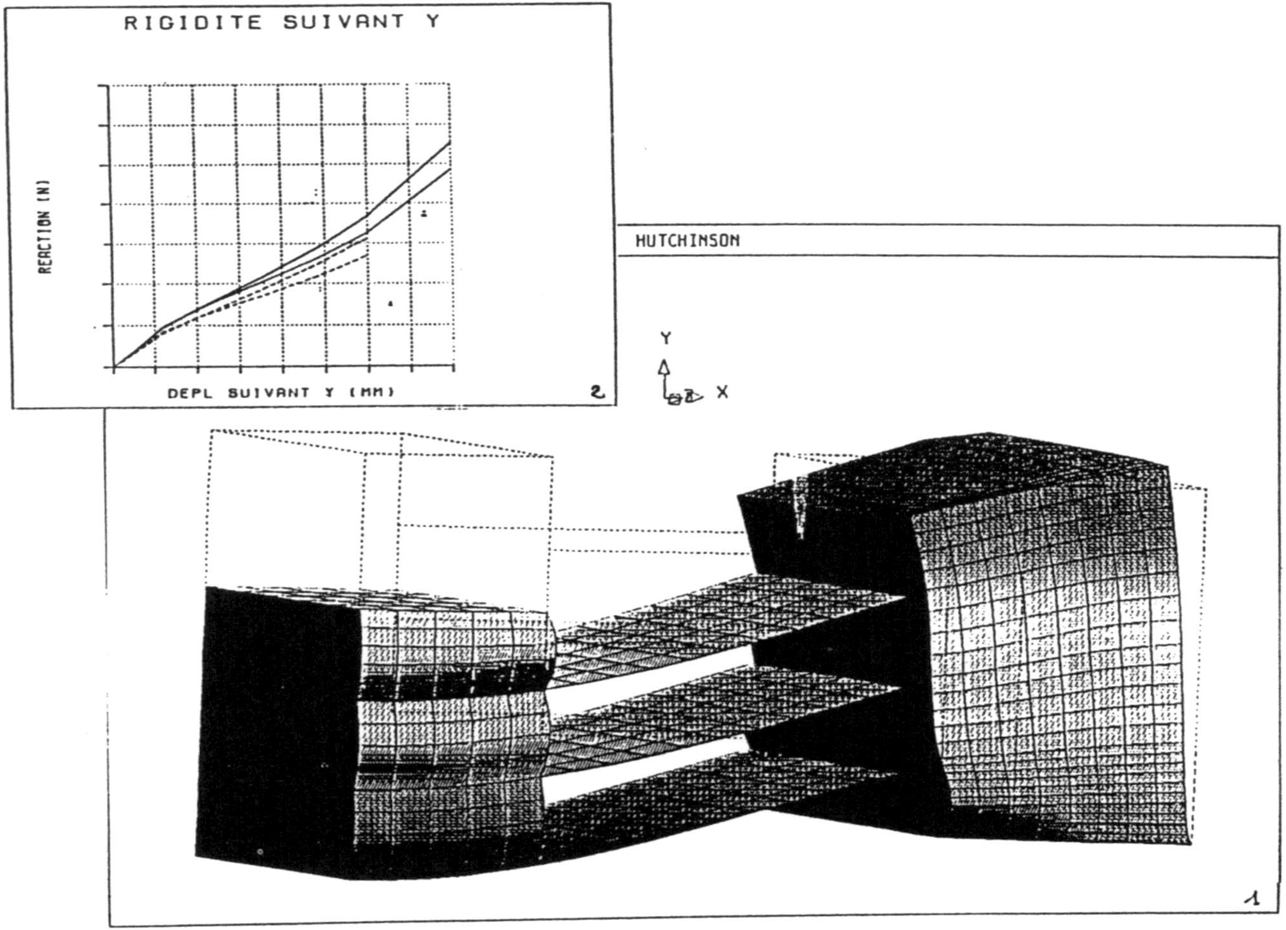
RIGIDITE SUIVANT Y
REACTION (N)
DEPL SUIVANT Y (MM)
2
HUTCHINSON
Y
X
1

FIGURE A

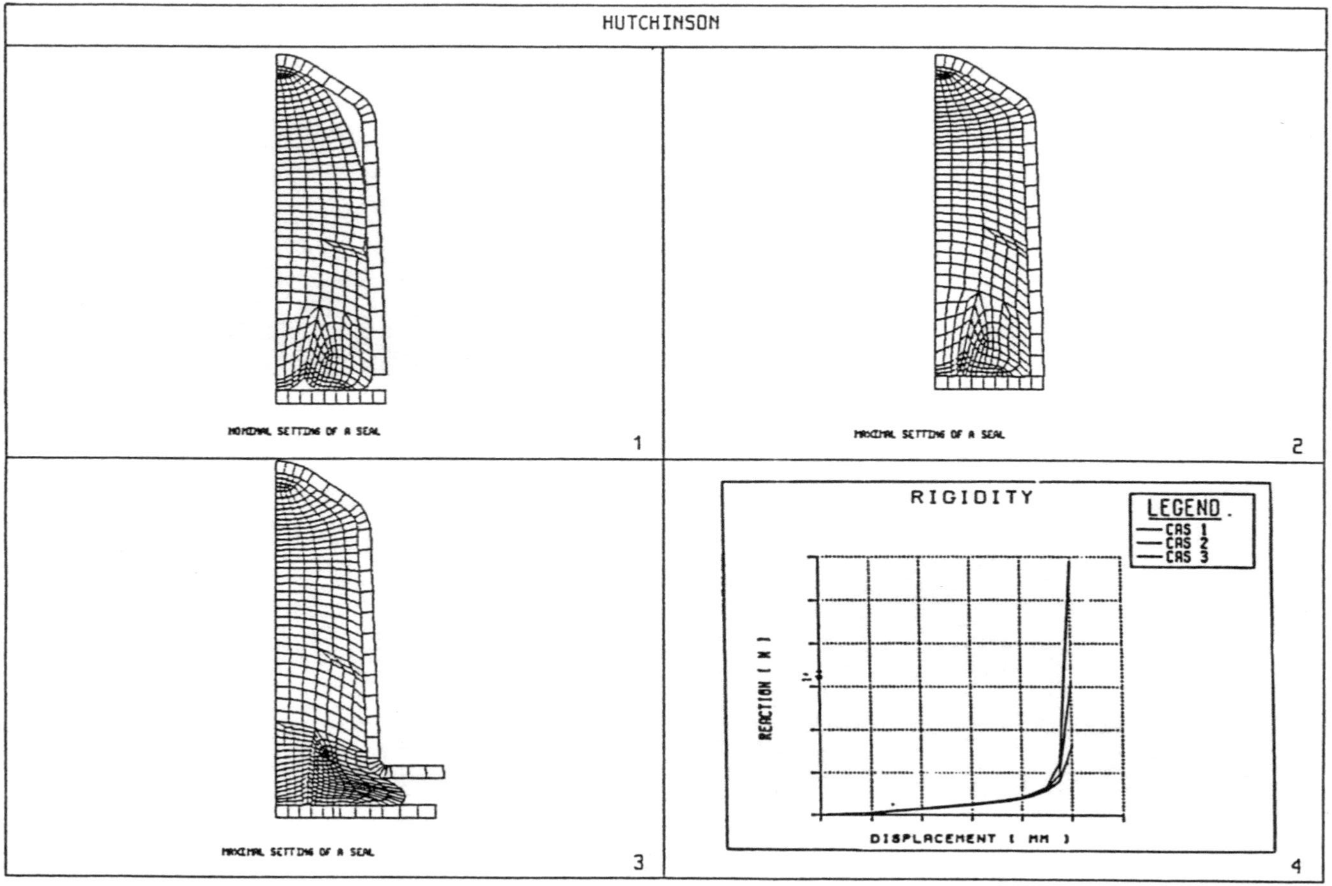
HUTCHINSON
NOMINAL SETTING OF A SEAL
1
MAXIMAL SETTING OF A SEAL
2
MAXIMAL SETTING OF A SEAL
3
RIGIDITY
LEGEND.
CAS 1
CAS 2
CAS 3
REACTION (N)
DISPLACEMENT (MM)
4

FIGURE 8

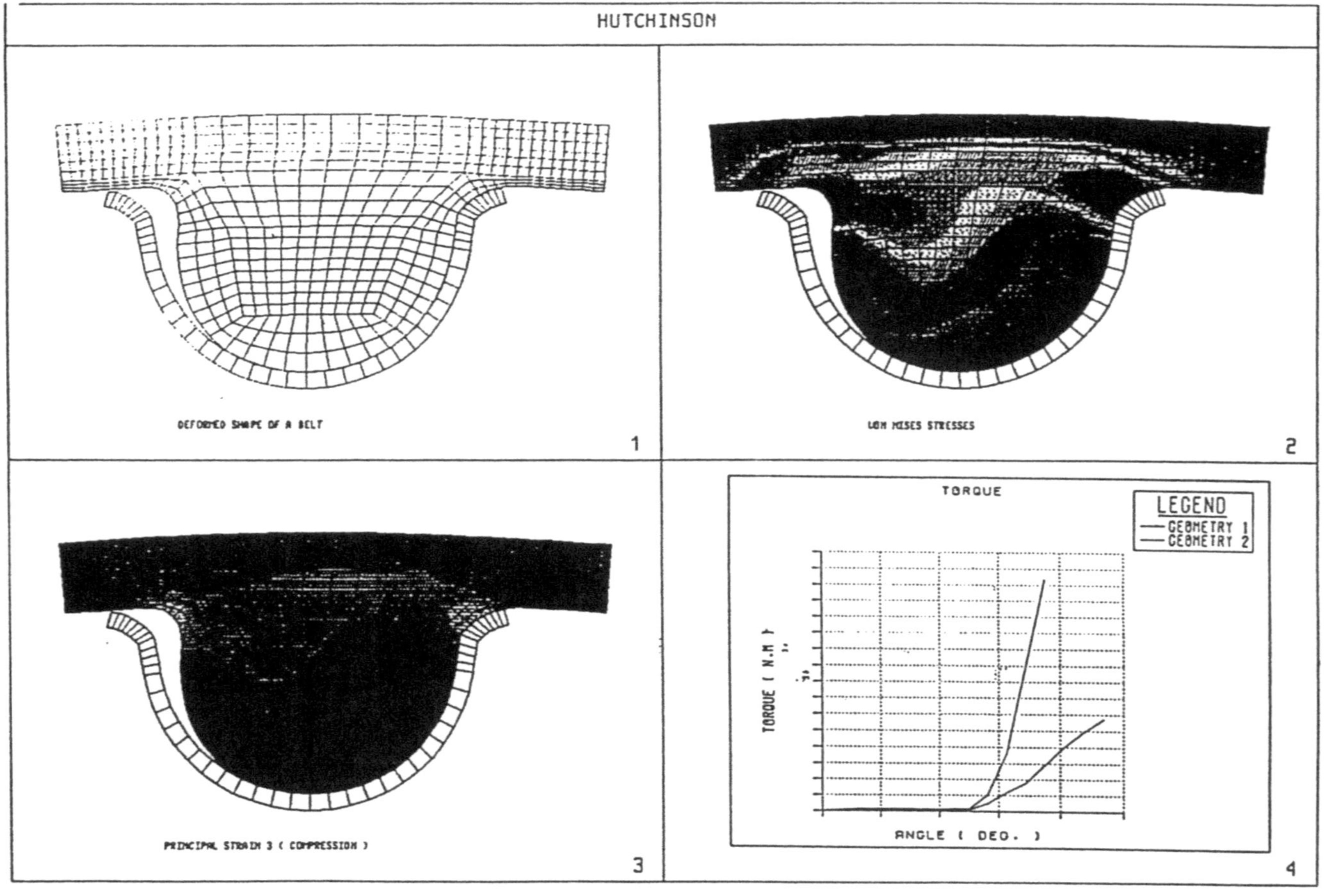
HUTCHINSON
DEFORMED SHAPE OF A BELT
1
VON MISES STRESSES
2
PRINCIPAL STRAIN 3 (COMPRESSION)
3
TORQUE
LEGEND
GEOMETRY 1
GEOMETRY 2
TORQUE (N.M)
ANGLE (DEG.)
4

FIGURE C

HUTCHINSON

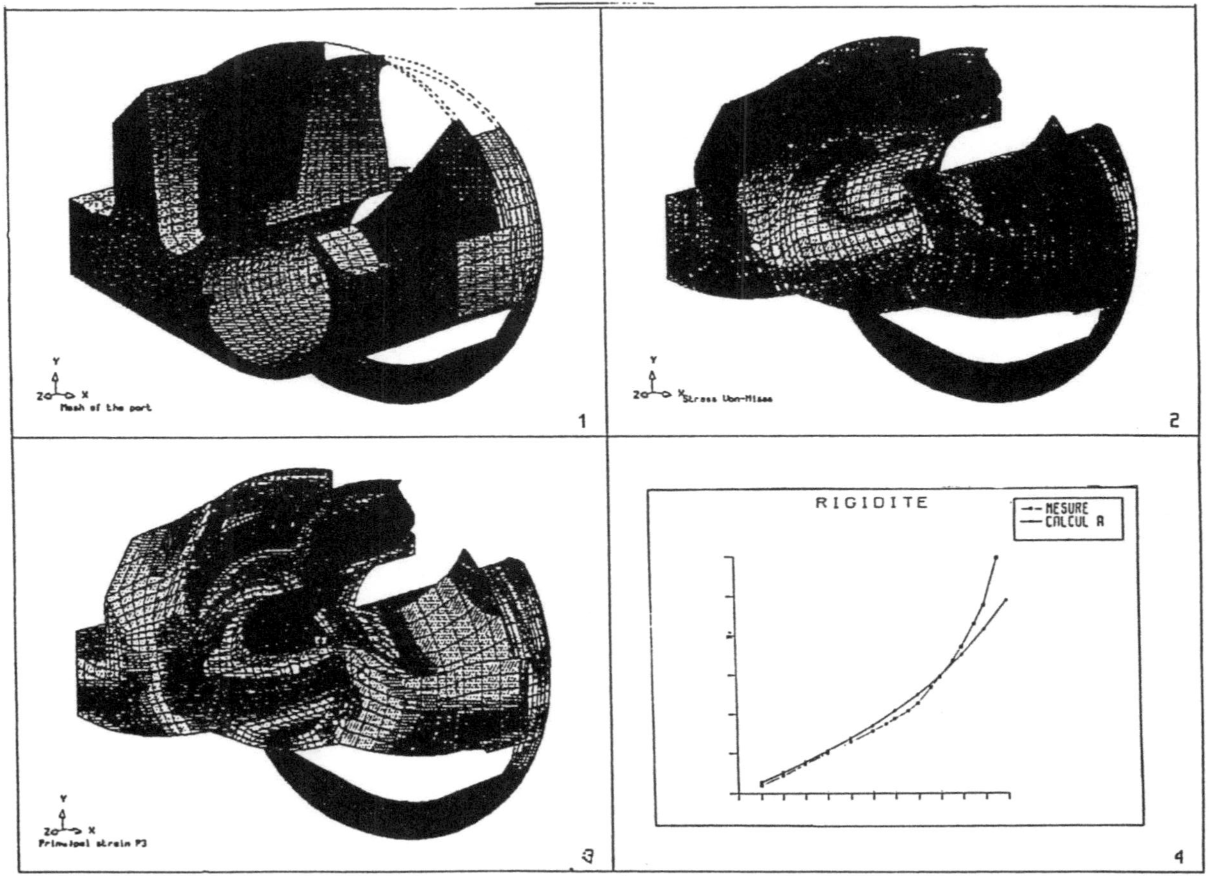

FIGURE D

Crashworthiness

Recent Trends and Developments of Crashworthiness Simulation Methodologies and Their Integration into the Industrial Vehicle Design Cycle

E. Haug, H. Charlier, J. Clinckemaillie, E. Di Pasquale, O. Fort, D. Lasry, G. Milcent, X. Ni
Engineering Systems International S.A., 20 rue Saarinen, 94578 Rungis-Cedex (France)

A.K. Pickett, R. Hoffmann
Engineering System International GmbH, Frankfurter Str. 13-15, 6236 Eschborn (WG)

SUMMARY

Crashworthiness simulation has become an accepted and even indespensable design tool in the development of transport vehicles. Further developments of numerical models, and efforts of integration into the industrial design cycle are still happening at an accelerated pace. Numerical model developments concern basic algorithms, specific user friendly options, new materials and new applications. New applications of crashworthiness codes are in the field of occupant safety and sheet metal stamping. This survey paper discusses such development trends and it illustrates the results on simple benchmarks and on full scale industrial examples.

INTRODUCTION

Crashworthiness simulation techniques for vehicle bodies made of thinwalled soft steel assemblies have now reached a level of maturity that permits their beneficial integration into the industrial vehicle design cycle, from the concept stage of design up to the prediction of the vehicle performance in legal crash tests. The winning spatial discretization technique is the underintegrated C^0-Mindlin type thin shell finite element approach for the simulation of the crash behaviour of thinwalled steel parts, which, for example, is backed by nonlinear beam, bar and solid models of gear, accessories, paddings and legal impact barriers. The explicit central difference time integration scheme has proven far more economical and superior to implicit integration techniques for large crash models with up to 30 000 and more shells. Von Mises plasticity material models, strain rate models and contact algorithms have been developed to treat nonlinear steel behaviour and contact effects with precision and computational economy.

New material models, such as for the description of layered glass, aluminum and composite behaviour are currently developed to complete the simulation panoply of crashworthy materials. Recent development efforts center around passive occupant restraint simulation techniques, such as airbags, belts, paddings and kneebolsters. The legal injury coefficients of occupants will be assessed via simpler or more elaborate numerical simulation models of occupant surrogates, such as rigid body linkage and deformable finite element dummy models. Contact algorithms for added efficiency and precision and for extremely complex folded airbag simulation events are under development. Automatic meshing and adaptive remeshing techniques are presently considered for the reduction in manpower and time delay for discrete model generation. Finally, workstation dedicated interactive software ("crash station"), including CAD interfaces for the efficient integration of the modern simulation techniques into the industrial design cycle and data base structures for local and interdepartmental communication, storage and retrieval of simulation results, experimental results, legal requirements and accumulated knowledge presently experience a surge of widespread interest and development. The paper tries to highlight and outline some of these trends and development activities from ESI's point of view, and to indicate preliminary results on illustrative examples.

COMPUTER INTEGRATED ENGINEERING

CARE system. With the broadening use of numerical simulation techniques, one of the most pressing issues becomes their efficient and structured integration into the industrial design and manufacturing cycle. Figure 1 symbolizes this trend towards CIE (computer integrated engineering), where a system of communication data bases and languages is used within a production industry to create a link between CAD (computer aided design) activities, CAE (computer aided engineering) activities, laboratory testing activities and CAM (computer aided manufacturing) activities. The link between these a priori disconnected activities can be established in part, for example, by ESI's CARE system (Computer Aided Results for Engineers [1]), that encorporates an interactive software, a data base management system (example : DAMES) and an information retrieval system (example : SPIRIT) for running simulation programs on EWS (engineering workstations), based on CAD data, and that obtains, displays, compares, processes, stores and retrieves and reports experimental and numerical result data for the final purpose of industrial manufaction via screen menus. At no point the user must access job control commands nor handle ASCII files, which is taken care of by the CARE software.

CRASH/CARE Station. Figure 2 gives an overview of a two level version of the CARE system that is adapted to exploitations of the PAM-CRASH/EWS code on engineering work stations. The higher CARE level contains functions like help, monitoring, on-line user's manuals, input preparation check list, tutorial for beginners, standard exploitation, extraction and comparison of results between runs and storage in the data base, automatic and fast comparison between plot time histories and technical report writing. The lower level of the CRASH/CARE Station software consists in the potentially stand-alone interactive PAM-CRASH/EWS pre, post and exploitation module. This module is activated from the higher CARE level and can accept and translate into PAM-CRASH format mesh data from CAD meshers such as SUPERTAB/ CAEDS, PATRAN, or from FE codes like NASTRAN and ANSYS. Next, it can complete PAM-CRASH inputs by adding complementary information such as contacts, material properties (from a material data base), and it can manipulate meshes by remeshing operations, folding operations for airbags, dummy positioning operations, etc. The module will then execute PAM-CRASH/ PAM-SAFE, including composite and occupant safety options (with airbags, belts, coupled dummies from MADYMO or CAL3D, FE dummies). Finally, the module will post-process the results via standard result display, graphic animation and video outputs.

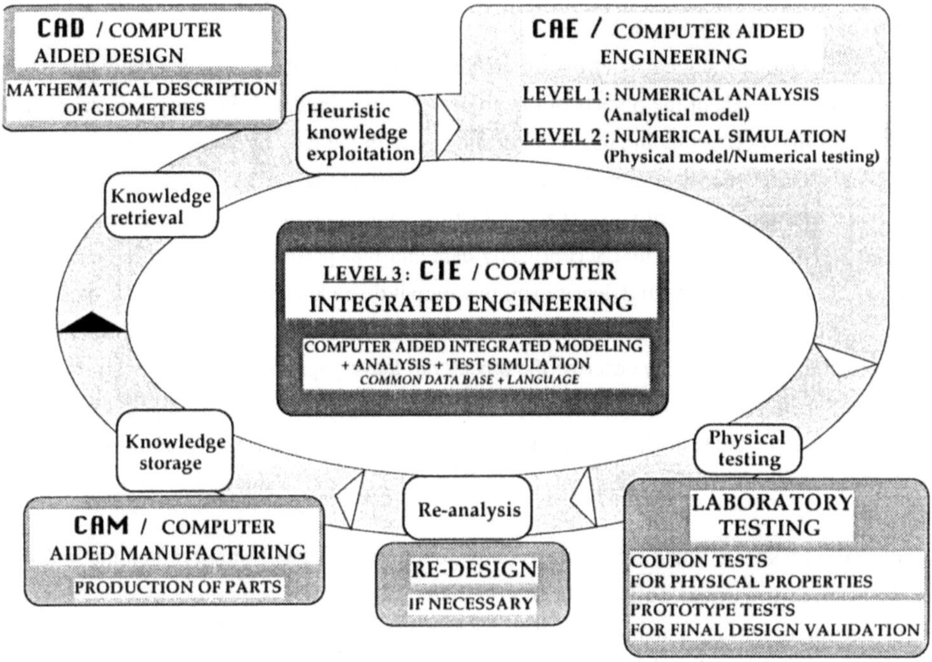

Fig. 1 : Computer Integrated Engineering (CIE)

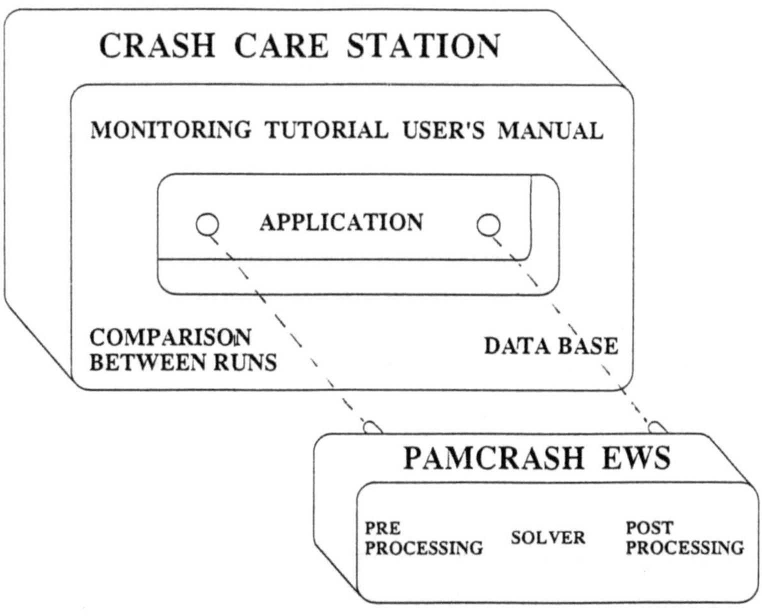

Fig. 2 : CRASH-CARE Engineering Work Station

CAD MESHING INTERFACES

CAD data. One of the greatest efforts in numerical simulation goes into the establishment of mathematical models of the objects of the processes to be simulated. In many industries such objects are first defined by the industrial designers, with the help of commercial or domestic CAD software, in the form of mathematical curves and surfaces. The so defined geometries must then be transformed into discretized numerical simulation models. This can be done with special software interfaces that accept CAD data and that permit to generate finite element meshes of the desired kind and density and that output discrete nodal coordinates and element connectivities. One of the difficulties for generating numerical models from CAD data is that they often do not completely describe the object, but leave gaps (e.g., omitting small radii) or have overlapping surfaces that are of no bother to the designer, but need special consideration by the numerical analyst. Such defaults must be overcome by the interface programs that actually are under intense development.

Automatic meshers. Much effort has recently gone into the development of automatic finite element mesh generation software from CAD data. Such software permits to specify mesh densities along surface patch border lines and/or average mesh densities in surface or 3D domains, and it automatically generates surface meshes for shell models or solid meshes for solid models.

Mesh refinement software. A simpler variant of above scheme consists in automatic refinement of existing coarse meshes, such as needed when producing "crash meshes" from "NASTRAN meshes" in the automotive industry. Such software will permit to interactively select subregions within which the algorithm will subdivide existing elements, into smaller elements.

CRASH SIMULATIONS

Two seemingly opposite trends can be observed today for crash models : the trend for car manufacturers to keep refining global crash models and to add more details such as bumpers, power train, suspension gear, wheels, steering assemblies, radiators, windshields, engines accessories, batteries, containers, tanks and airbags, belts, seats and dummies, etc., on the one hand, and on the other hand the trend towards very simplified crash model manipulation during the early design phase that can give rough answers with little effort of manpower and CPU time.

Front and side impact simulations. Two examples may illustrate the trend towards more detailed large finite element models for passenger car crash simulations. The first example concerns a frontal crash simulation carried out by HYUNDAI with PAM-CRASH on a model that uses about 24 000 thin shell elements, Figure 3, Reference [2].

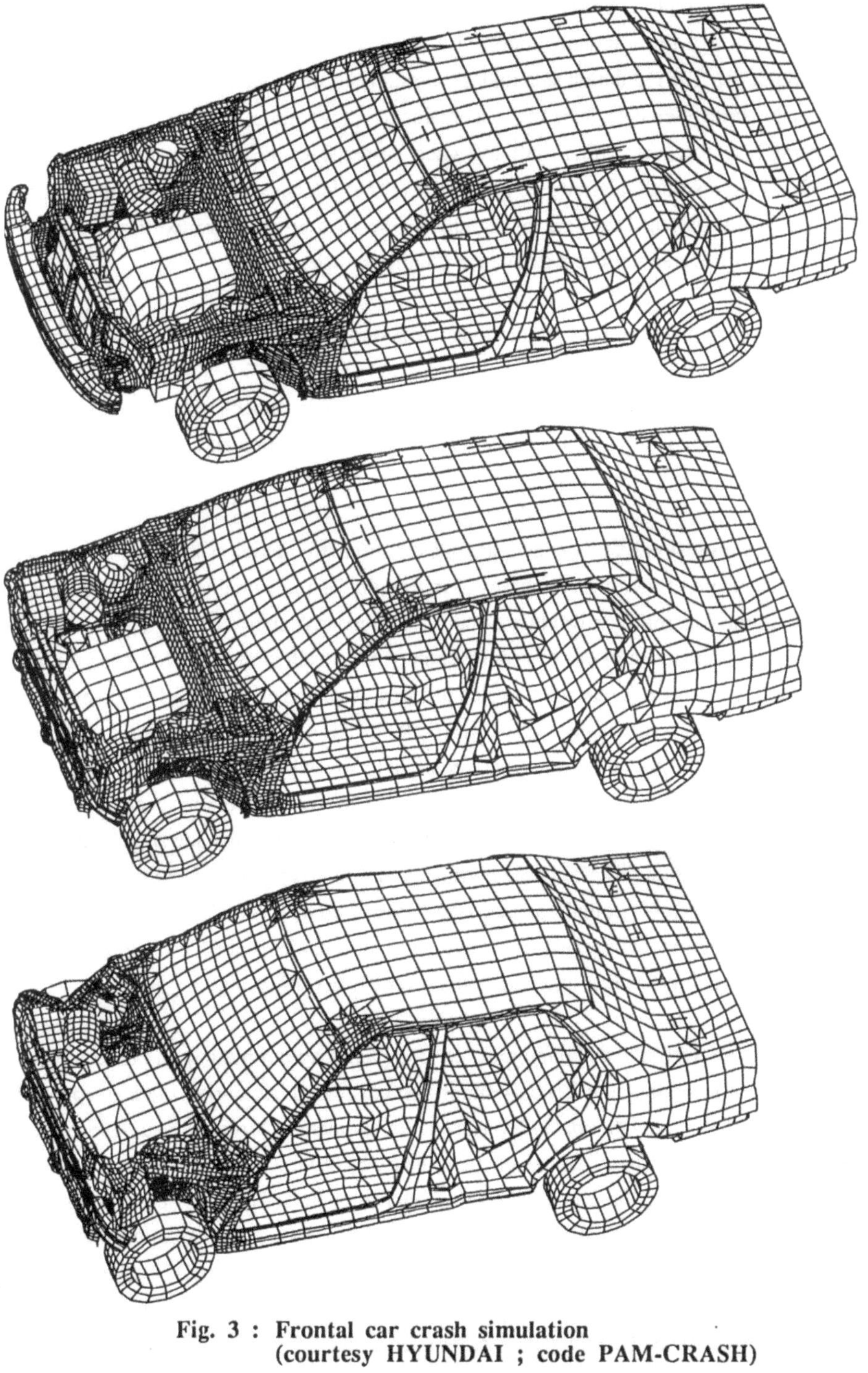

Fig. 3 : Frontal car crash simulation
(courtesy HYUNDAI ; code PAM-CRASH)

The second example is a side impact barrier test simulation performed by BMW with PAM-CRASH on a finite element model with about 28 000 thin shells and beams, Figure 4, Reference [3]. The regulation side impact barrier has been modeled with a moderate number of crushable brick elements. This model contains the power train, including beam models of the steering assembly, the cardan shaft and wheel suspensions, as well as detailed shell and brick models of the wheels and engine accessories. The deformed shape of the model after side impact is also shown. The same model may also serve after modular modifications of the mesh density for front and rear crash simulation ("unified crash model").

Beside front and side impact simulations numerous other types of passenger car crash simulations have been carried out recently and reported in the open literature, such as rear impacts, oblique and offset front impacts, car-to-car impacts, pole impacts, etc. All these simulations may be considered variants of the basic frontal crash simulation and need not to be discussed further.

Rollover simulation. A passenger car rollover simulation requires at least one new feature : the management of intermittent phases of free flight. The potentially very CPU time intensive long duration of a rollover event (minimum : 2-3 seconds), must be fully integrated to capture the important episodes of car-to-ground contact. One such simulation has been reported [4], where a BMW passenger car rollover sled test was simulated on a relatively coarse FE model of about 4 650 shells and beams (ESI GmbH ; PAM-CRASH). The simulation code has been adapted for that purpose to simulate the free flight phases with the entire model transformed into a master rigid body, where only the six degrees of freedom of the center of gravity are integrated during flight. Upon detection of incipient ground contact, the rigid body model was partly transformed into a deformable FE model, that could respond by appropriate deformation to the brief phases of ground contact. Upon loss of contact, the deformable portions of the model were automatically transformed back into rigid bodies and added to the full car master rigid body model for explicit time integration over the next flight phase.

"Concept car design". Although most car companies do not hesitate to fabricate very large scale FE crash models, the issue of simplified crash analysis during the conceptual design phase keeps being brought up. One consortium of car manufacturers, for example, has fostered development of the "superfolding" element theory by Wierzbicki and Abramovicz [5], that is the basis of a software that can be run on personal computers. The axial deformation version of this element for the analytical evaluation of axial crushing forces of arbitrary geometry prismatic thinwalled box beam sections has been built into a test version of the PAM-CRASH code, and it is planned to incorporate an advanced version of this "element" that includes load excentricity and combined axial and bending collapse, as soon as the software is available.

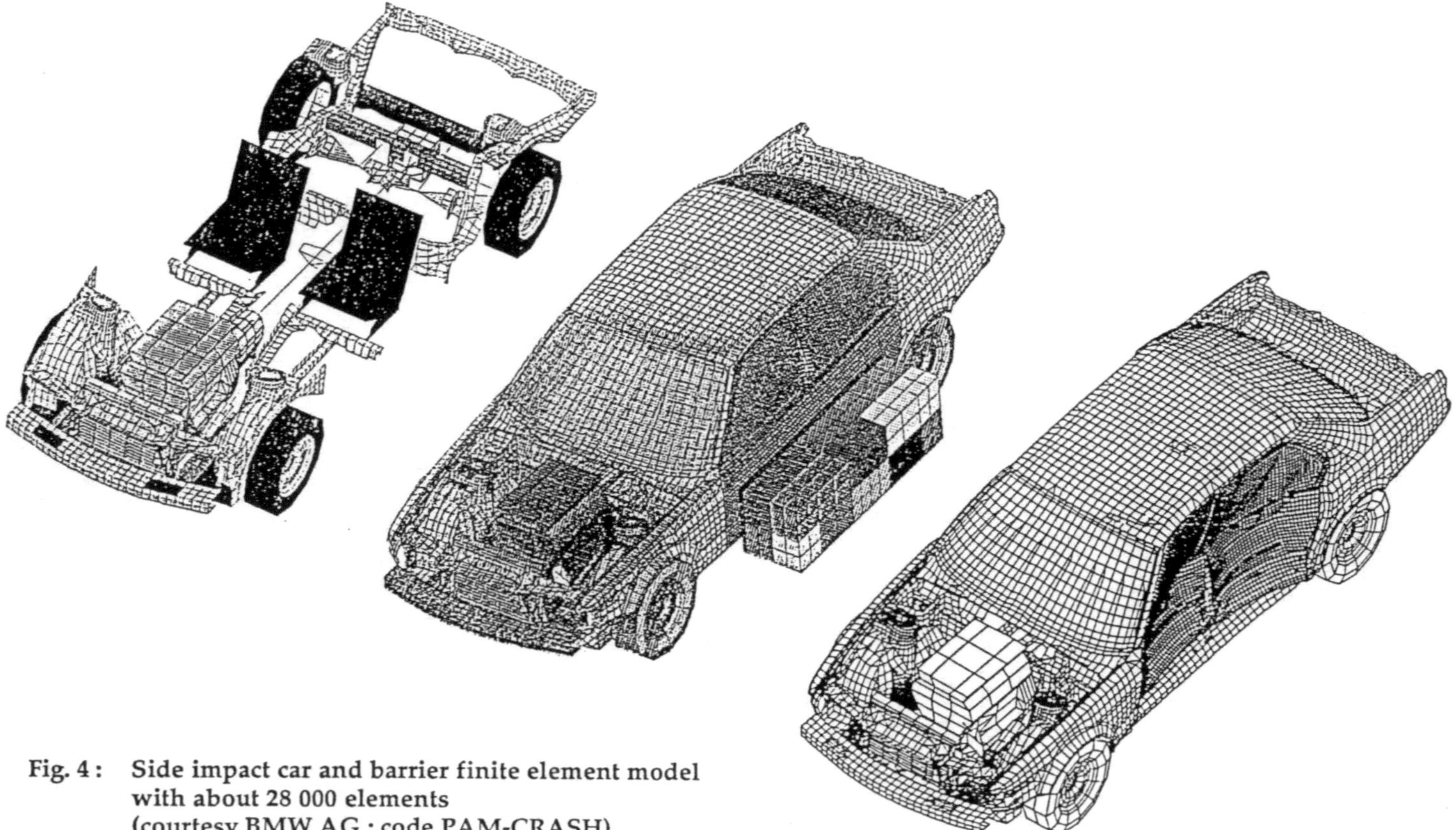

Fig. 4 : Side impact car and barrier finite element model with about 28 000 elements (courtesy BMW AG ; code PAM-CRASH)

Other popular means to perform simplified crash analyses are nonlinear beam, bar, six degree of freedom spring and dashpot elements which allow the user to "lump" known nonlinear component responses. It has also be found very effective to calibrate, e.g., the yield stress of regular finite elements in regions of coarse mesh densities such as to absorb the same amount of crash energy than a fine mesh of the same component or of a tested component would absorb ("macro" elements). The added advantage of the latter method is that the component topology can be preserved by the equivalent coarse mesh of the considered component, which increases plot quality and the chance for acceptable response in the case of unforeseen load excentricities. Several such concepts are under active development and may be incorporated into existing crash codes.

NEW MATERIALS

Crashworthiness investigations of structures made from material other than car body steels, such as aluminum and composites, experience a considerable upsurge in interest.

Aluminum. Guided vehicles such as subway, train and high speed train cars are increasingly built from aluminum alloys. Recently, aluminum appears to experience major break throughs as an alternative material for passenger cars. Concerning crashworthiness events, components made from aluminum may either retain their structural integrity, display surface cracks or through-the-thickness cracks, or the component may completely loose its structural integrity, by breaking into a multitude of separate fragments. Behaving initially like an elasto-plastic material, aluminum will behave like a brittle fracturing material beyond a certain limit of plastic strain. This dual behaviour can be simulated approximately by introducing a branch of negative strain hardening, down to almost zero residual stress, into a convential von Mises type elasto-plastic model. More accurately, the fracture behaviour can also be simulated by damaging the elasto-plastic stresses after having exceeded a threshold strain.

Composites. Recently [6] a complete frontal crash simulation of a passenger car cabin from a two seater prototype sportscar, made 100 % of carbon-Kevlar-aramid honeycomb sandwich material, has been performed at ESI in a joint study with TONEN Corporation, that was sponsored by MITI (Ministry of International Trade and Industry, Japan). This cabin was tested at INRETS (Lyon). The cabin wall material consisted of multilayered and multimaterial composite stackups with (0°, 90°) cross plies of TONEN high strength/high modulus FT500 pitch based carbon fiber material, and with Dupont aramid fiber KEVLAR49 (0°*90°) and TORAY PAN based T300 carbon (+45°*−45°) cloth plies, Figure 5. The sandwich facings were modeled with a multilayered elastic-fracturing Mindlin type thin shell finite element, built into PAM-CRASH, where the centers of the physical plies are assigned to shell thickness integration points with the orthotropic material properties of the corresponding ply material, including fracturing behaviour. The sandwich core was modeled using nonlinear orthotropic eight node brick solid elements and the entire sandwich walls were modeled by stacking two multilayered thin shells to the two opposite faces of the brick elements.

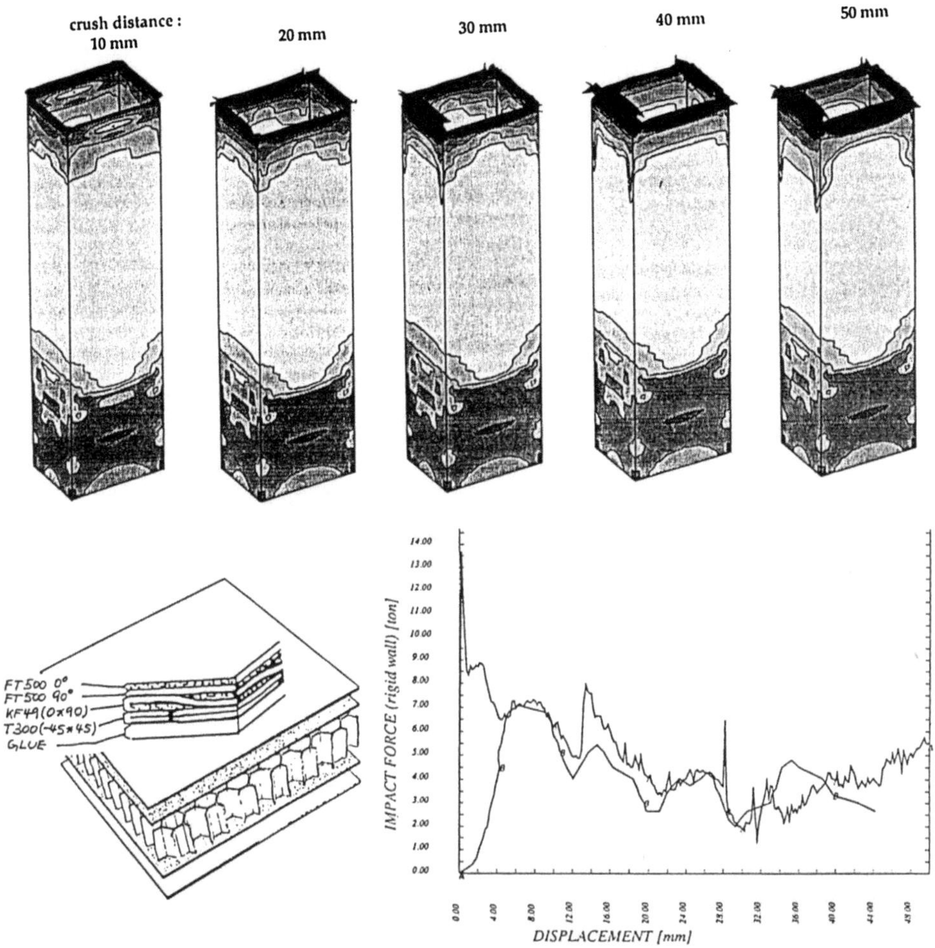

Fig. 5 : **Composite sandwich box beam axial crush test simulation (after TONEN ; code PAM-CRASH)**

Figure 5 summarizes the results of a multilayered and multimaterial sandwich wall box beam axial crush simulation with superimposed average damage contours on the outer facing, with an excellent comparison of the numerically obtained impact force-displacement curve with the experimentally obtained curve. These results have been obtained after calibrating all involved material properties on standard coupon tests, and by validation of these calibrations on the measured test results. Figure 6 shows the deformed shapes of a 6500 sandwich element (i.e. 6500 bricks + 13000 shells) plus 2000 thin shell element model (total : 6500 bricks + 15000 shells) of the prototype car cabin upon frontal impact onto a rigid wall. The overall composite plate damage and the damage of the sandwich core material, as well as the predicted crash distance were in good agreement with the experiments.

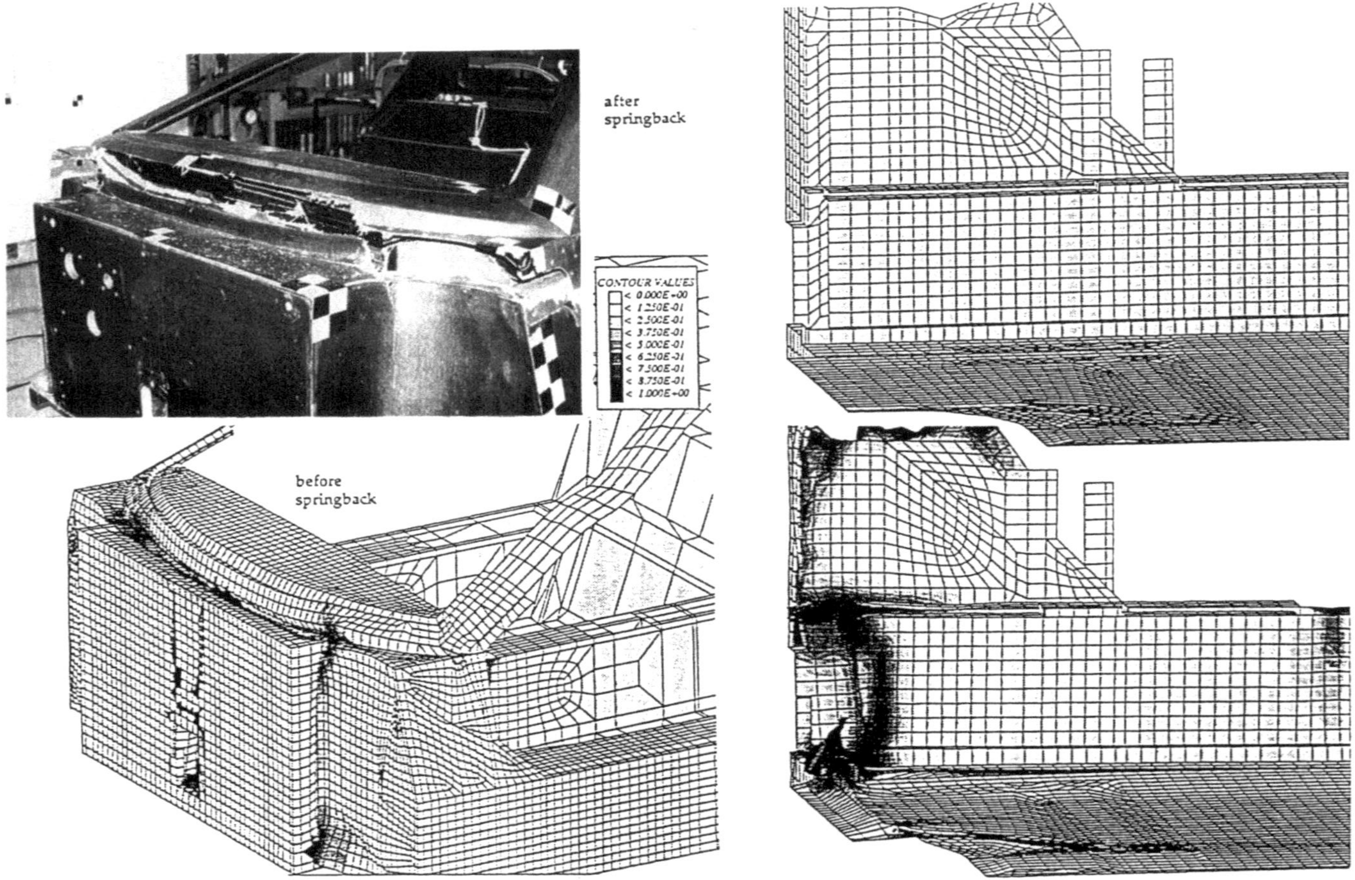

Fig. 6 : Composite prototype passenger car cabin frontal crash simulation (after TONEN ; code PAM-CRASH)

OCCUPANT SAFETY

The ultimate goal of passenger car crashworthiness investigations is to ascertain or to provide occupant safety [7]. Considerable efforts have gone into the simulation of occupant safety via airbags, dummy models, seat belts, seats, etc. Some recent trends and developments in this area are outlined below.

Airbags. Airbags are passive occupant restraint systems made of flexible fabric that are stored inside sealed chambers and that are inflated by gases developed by chemical reactions, triggered in the case of a car accident via sensors that respond to sudden decelerations. Their numerical simulation with finite element models of an initially folded and then deployed fabric membrane and with gasdynamic equations describing the interrelationships between pressure, volume, mass and temperature of the enveloped gas, and the simulation of their interaction with occupant surrogates (dummies) and with obstacles inside the passenger compartment, have now attained a certain degree of feasibility, but still require further development. While the description of the fabric is satisfactory, see, for example, the airbag membrane element of PAM-CRASH with isotropic coating plus visco-elastic straight or nonlinear woven fiber reinforcements (warp/weft kinematic interaction and nonlinear fiber stretch model), tension-only option (convenient absorption of fabric wrinkles without mesh undulations), finite shear deformation angles between weave directions, memory of intrinsic flat layout surface metric to be used for folded meshes (compensation of inevitable warp due to folding of complex airbags), performant contact algorithms, etc., the description of the gasdynamic phenomena inside the chamber, as well as of gas outflow through vents, leaking fabric and seams is under continued development at present. While some groups prefer to directly discretize the gas equations, ESI developed a multi-chamber airbag that is based on simple gasdynamic equations such as proposed by Wang and Nefske) but that exploits the fact that an initially folded airbag forms chambers or pockets, delimited in a natural way by the folding lines of the folded bags, Reference [8].

Airbag/dummy interactions. Figure 7 is a generic picture made by ISUZU for demonstration purposes, where a folded driver side airbag and a folded passenger side airbag are fixed simultaneously and interact with two dummies in the driver and passenger seat. This benchmark type analysis has been performed with the standard PAM-SAFE models on a SILICON GRAPHICS workstation, in conjunctin with the MADYMO code (dummy models), using the PAM-CVS coupling software. The image demonstrates the fact that one or several airbag/dummy interaction simulations are well within the present state of the art.

Seat belt models. Figure 8 shows the convenient application of a seat belt model, initially fixed at one point on a dummy chest and after automatic positioning on the chest. The nodes of the user beam model can slide (with friction) on the dummy chest model.

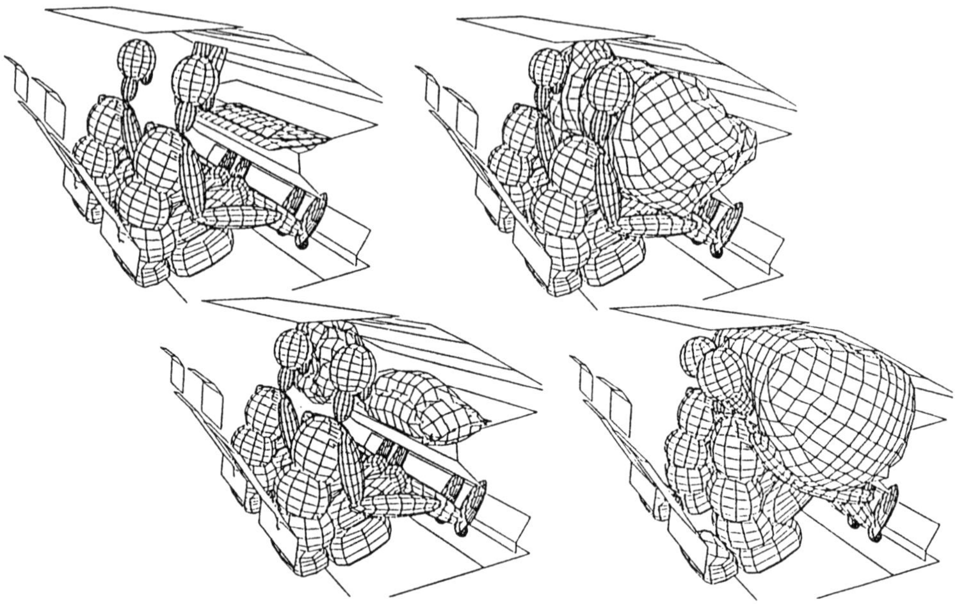

Fig. 7 : Driver and passenger airbag inflation simulation with dummy interactions (after ISUZU ; code PAM-SAFE)

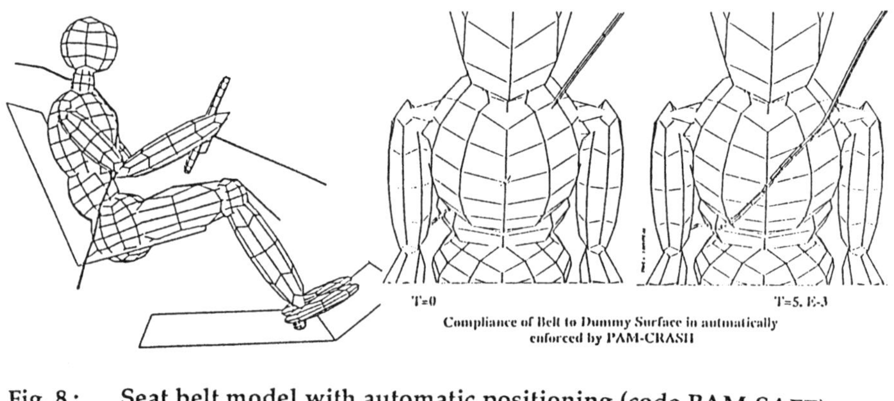

Fig. 8 : Seat belt model with automatic positioning (code PAM-SAFE)

Deformable dummies. Considerable development efforts are expended today for generating more realistic dummy models than the standard rigid linkage models built into the CAL3D and MADYMO crash victim simulation softwares. One such effort carried out jointly by Wayne State University and ESI [9] is to produce a HYBRID III front impact occupant surrogate model that has deformable parts, modeled with deformable finite element subassemblies, while other portions may be modeled as rigid linkages, connected by joint elements. Figure 9 shows the finite element thorax submodel of the HYBRID III dummy with thin shells, bricks and rigid parts, including a massive pendulum (23.5 kg) that moves horizontally and impacts the chest model at a given initial velocity of 6.7 m/s, and Figure 10 are deformed shapes obtained from the simulation of a laboratory pendulum impact test. The agreement of this deformable chest submodel with a full scale dummy test is quite good, but shows some residual oscillations after impact that will vanish when the full dummy test is simulated and when dispersive elements, such as the dummies "vest", are included in the finite element model.

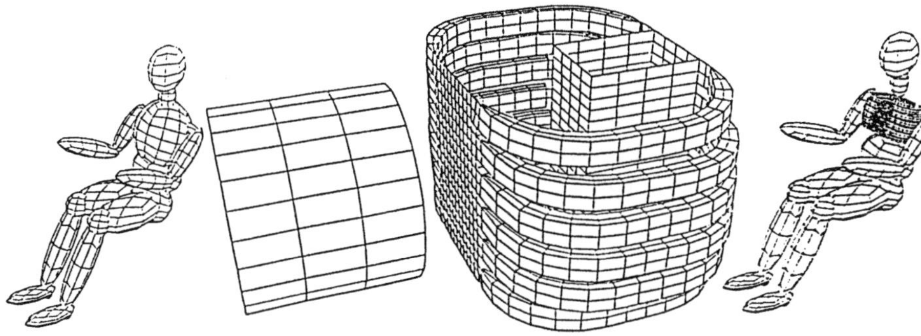

Fig. 9 : Deformable HYBRID III finite element thorax submodel (after Wayne State University ; code PAM-SAFE)

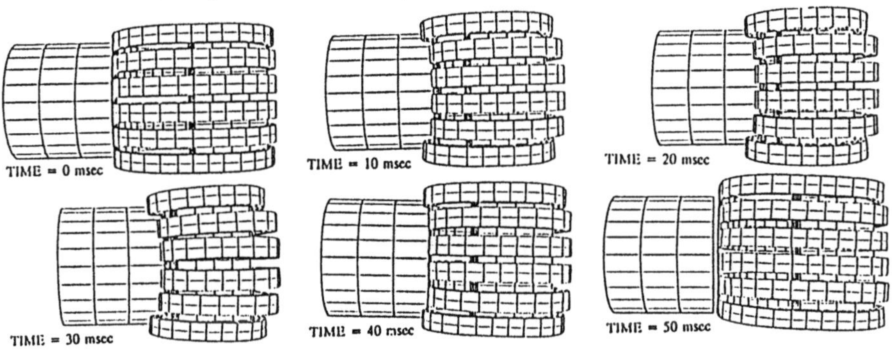

Fig. 10 : Deformable HYBRID III pendulum test simulation thorax deformations (after Wayne State University ; code PAM-SAFE)

STAMPING SIMULATION

Although not directly related to the crashworthiness issue, stamping simulation merits mention at this place, because it can utilize the same basic numerical simulation techniques than are used for crashworthiness simulations. For technological reasons, however, stamping simulation requires specific extensions to anisotropic plasticity models, accurate and strain amplitude dependent strain rate models, consideration of shell thickness changes, nonlinear pressure and velocity dependent friction laws, very precise contact algorithms and reliable predictions for sheet metal tearing and wrinkling. Moreover, quasi-static elastic spring-back calculations are needed in order to assess the final shape of a stamped part after retraction of the stamping tools (punch, die and blankholder). Above special features are actually developed at ESI in the context of a BRITE-EURAM project [10].

Adaptive meshing. A particularly interesting numerical feature for stamping simulations may prove to be automatic adaptive finite element mesh refinement and unrefinement algorithms in combination with subcycling algorithms, when the metal sheets are pulled over sharp tool edges. Such schemes are under development and their application will contribute to simulation precision and CPU time efficiency.

Rectangular cup stamping simulation. Figure 11, finally, gives the results of a stamping simulation with a prototype stamping simulation code, PAM-STAMP, for a rectangular cup, made of ductile car body steel. Details on stamping simulation additions to the explicit analysis code can be found in Reference [10].

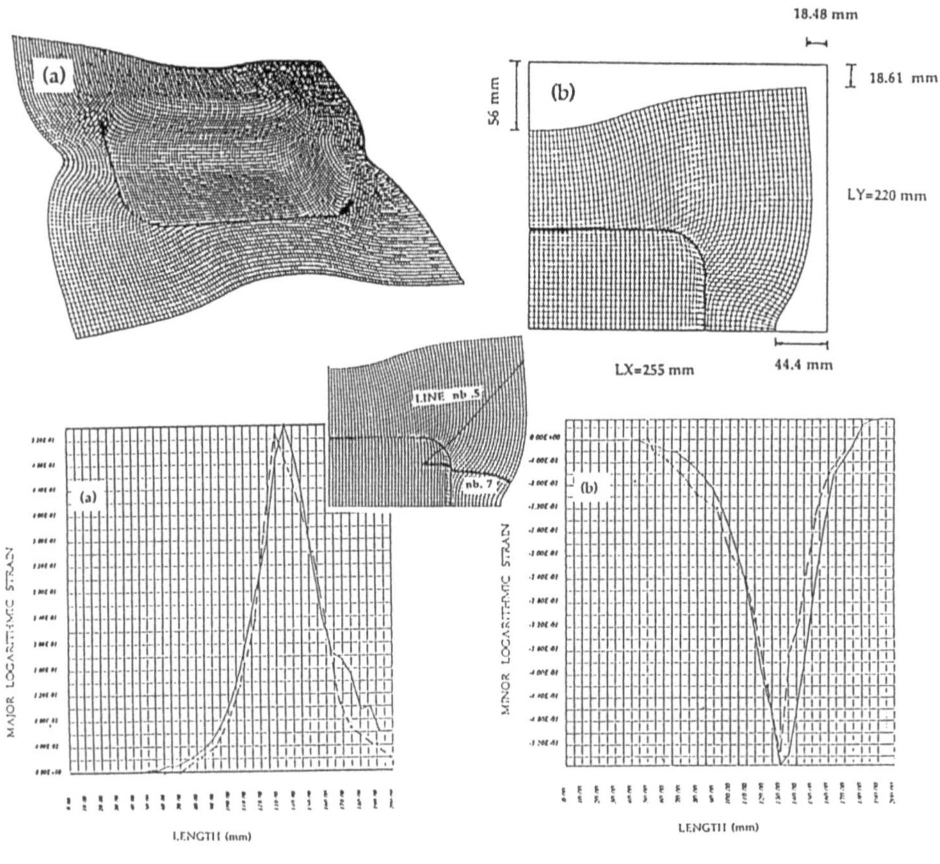

Fig. 11 : Stamping simulation of a rectangular cup (after FIAT ; code PAM-STAMP)

ACKNOWLEDGEMENTS

The authors are indepted to so many collegues and engineers from the industry that they do not make an attempt to mention any names. They express their gratitude and esteem to all these contributors.

REFERENCES

Only the direct references on which some of the presented material is based are given here. Within these references, the interested reader may find a host of further references on the discussed subjects.

[1] Dubois, J., Valent, R., Gregis, J.L., Carlier, P., Klein, M. and Kreis, A., "Exploitation of Data Management Facilities for Finite Element Analysis in Aerospace Structural Engineering Applications", Intl. Conference 'Spacecraft Structures and Mechanical Testing', ESA-ESTEC, Noordwijk, The Netherlands, October 19-21, 1988.

[2] Park, K.H. and Cho, H.S., HYUNDAI, private communication in the framework of a joint study with ESI on the crash simulation of a HYUNDAI passenger car, August 1991.

[3] Höck, H.G., "Accident Simulation Experiences at BMW", 1-st PAM USER's SEMINAR and Numerical Simulation Workshop (Ed. ESI-GmbH, Frankfurter Str. 13-15, 6236 Eschborn, Germany), Bad Soden, June 11-13, 1990.

[4] Pickett, A.K., Höck, H.G., Poth, A. and Schrepfer, W., "Crashworthiness Analysis of a Full Automotive Rollover Test Using a Mixed Rigid Body and Explicit Finite Element Approach", VDI Conference 'Berechnung im Automobilbau', Würzburg, Germany, October 2-3, 1990.

[5] Abramowicz, W., "Superfolding Element in PAM-CRASH", 1-st PAM USER's SEMINAR and Numerical Simulation Workshop (Ed. ESI-GmbH, Frankfurter Str. 13-15, 6236 Eschborn, Germany), Bad Soden, June 11-13, 1990.

[6] Haug, E., Fort, O., Tramecon, A., Watanabe, M. and Nakada, I., "Numerical Crashworthiness Simulation of Automotive Structures and Components made of Continuous Fiber Reinforced Composite and Sandwich Assemblies", SAE910152, Intl. Congress and Exhibition, Detroit, Michigan, February 25-March 1, 1991.

[7] Lasry, D., Hoffmann, R. and Protard, J.B., "Numerical Simulation of Fully Folded Airbags and their Interaction with Occupants with PAM-SAFE", SAE Intl. Congress and Exhibition, Detroit, Michigan, February 25-March 1, 1991.

[8] Haug, E., Protard, J.B., Milcent, G., Herren, A. and Brunner, O., "The Numerical Simulation of the Inflation Process of Space Rigidized Antenna Structures", Intl. Conference on Spacecraft Structures and Mechanical Testing, ESTEC, Noordwijk, The Netherlands, April 24-26, 1991.

[9] Hong Pan, "Finite Element Modelling of the HYBRID III Dummy Chest", Thesis for the Degree of Master of Science, Wayne State University, 1991.

[10] Haug, E., Di Pasquale, E., Pickett, A.K. and Ulrich, D., "Industrial Sheet Metal Forming Simulation using Explicit Finite Element Methods", Intl. VDI Conference with Workshop 'FE-simulation of 3D Sheet Metal Forming Processes in Automotive Industry', Zürich, May 14-16, 1991.

Quasistatic Seat Belt Anchorage Analysis with Explicit Time Integration

Georg Giazitzis , Ford Werke AG - Köln

Abstract

An approach will be presented to use the explicit time integration (crash code) for quasistatic analyses. In order to verify this approach a seatbelt anchorage analysis was performed.
The stable time-increment for the explicit time integration procedure is proportional to the square root of the mass density. If the analysis can be handled quasistatically the mass density becomes a free parameter and can be increased to increase the stable integration time-increment reducing thereby calculation- and turnaround-time. Different time dependant loading characteristics have been analyzed and tested in order to minimize dynamic effects. The appropriate loading curve together with a mass density increase served to approach a quasistatic solution.
In this investigation the mass density value was increased by factors of 10 and 100. Results were assessed and compared to results using the original (unchanged) mass density and also to test results.
The analyses performed led to the conclusion that a quasistatic analysis is possible with an explicit integration programme and calculation time can be reduced considerably by increasing the mass density value and using a matching loading curve. The analysis is always stable and reliable, results obtained so far are conservative upper bounds.

Objectives

The objective of this investigation was to assess whether an explicit code (crash code) can be used to approach quasistatic solutions by increasing the mass density of the structure. The mass density can be regarded as a "free" parameter provided that no dynamic effects are involved. Since dynamic effects are introduced by the special loading, however, the scope of the mass density increase was to be defined.
Furthermore, it was to be analyzed how the inertia effects could be kept small with appropriate loading conditions and how the effect on results could be controlled.
The CPU-time reduction due to the mass density increase was to be related to the fact that results can be influenced by the mass variation.
The proposed method was to be verified on an analysis of a seat belt anchorage pull procedure. Data from a test was used to assess the method for its applicability and its reliability.

Background

Non-linear quasistatic problems which include material and geometrical non-linearities are normally tackled with implicit time integration methods allowing an accurate solution within an iterative solution scheme.
The algorithms are in most cases reliable, although this method appears to have some shortcomings depending on the degree of non-linearity in the task.
An iteration process in a 3-D analysis requires an update of the stiffness matrix, the geometrical matrix and in most cases the mass matrix, and increases therefore the calculation time considerably. This increase is even more drastic if the degree of non-linearity and the number of degrees of freedom (DOF) increase.
The contact can also play an important role for the reliability of the method. The contact zones have to be predefined in many implicit codes which of course is not always possible, since the deformation history of the structure is not known a priori. Depending on the formulations used (penalty method, Langrange multipliers or solver constraints) the contact algorithms are often not robust enough and do not converge for all kind of contact problems. In other words, the load cannot be applied completely.
To overcome these shortcomings other types of solution are being looked at.

Although the explicit time integration method has been established and well verified for shock-wave-like problems this method is being used in recent quasistatic applications like deep drawing processes [1]. Using a finite deformation velocity such a quasistatic process can be handled in an appropriate time-span without influencing certain results like deformation shape and thickness (but not stresses for instance).
The main idea in the method presented here is to increase the mass density by a certain amount and to increase herewith the stable integration time step. Since the stable integration time increment is proportional to the square root of the mass density, the mass density increase leads to the larger time increment already mentioned. This provides the possibility of regarding a wider time-span with less integration cycles. The possible wider time-span serves to approach a quasistatic solution provided that the mass density increase is moderate enough not to introduce too large inertia effects in conjunction with the loading characteristic. The loading characteristic is controlled by the load versus time curve which is to be analyzed and optimized for this purpose.
If the inertia effects can be kept sufficiently small the mass density becomes a "free" parameter within a certain range. In other words, the mass density could be increased in this range without influencing the results.
Since the dynamic effects cannot be avoided completely, however, limits occur for the scope of the mass density increase. In other words, the range of variation must be kept small in order to keep the deviation from the exact solution small. On the other hand, inertia effects can lead

to conservative displacement figures if for instance a deformation analysis is being performed.

All the finite element analyses have been performed with the explicit time integration programme PAMCRASH which is being distributed by ESI Germany.

Preliminary investigations

In a preliminary analysis the method of approaching a quasistatic solution by increasing the mass density and introducing an appropriate load versus time curve for the loading has been studied. The main aim of this preliminary study was to assess the proposed method in terms of reliability and applicability and to define the scope of the parameters which can influence the results.
The method has been applied to a deformation analysis of a square plate which was loaded perpendicular to the surface. The load was applied with two different time dependent load curves at the center of the plate. These time dependent loads were such that the load was linearly increased from 0 to 10 N in the first msec and afterwards increased from 10 to 500 N up to 20 msec or 30 msec then kept constant. Taking symmetrical boundary conditions into account one quarter of the plate was modelled (**figure 1**).

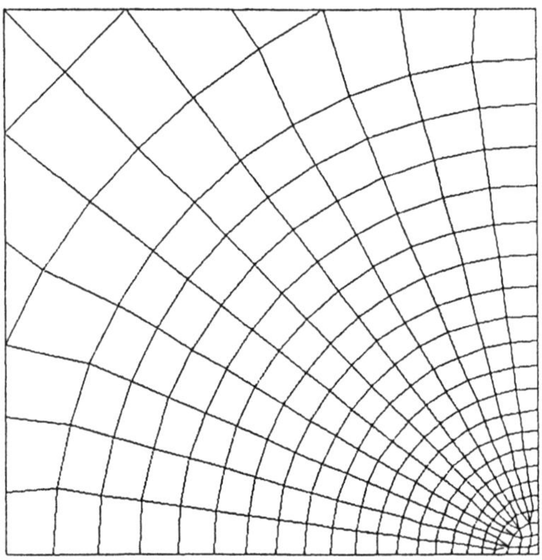

Figure 1: Quarter model of a plate

Fixed boundary conditions (fixed rotational degrees of freedom, fixed tranlational degrees of freedom) have been used for the outer boundaries.
The plate dimensions were **250 mm x 250 mm**. The thickness of the plate has been varied from **0.6 mm to 0.8 mm** in different runs influencing thereby the plate stiffness.
The material properties of a **ST04** steel have been used with an yield stress of **170 N/mm^2**.

Results of the preliminary investigations

Figures 2, 3 and 4 show some typical results due to the mass density variations with ρ_0 (initial density), **10 x** ρ_0, and **100 x** ρ_0 respectively.
Curve 1 in the legend refers to a loading curve having the larger slope up to the maximum value (**curve 2** with the smaller slope).

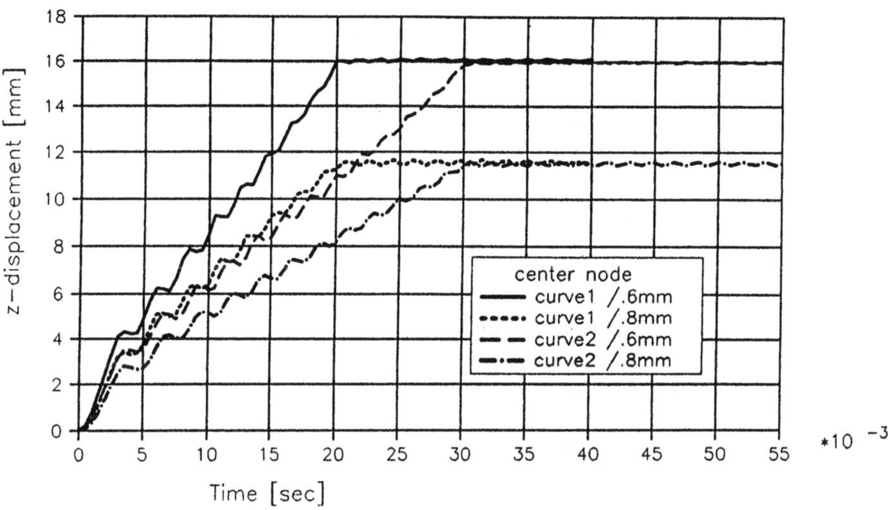

Figure 2: Square plate loaded perpendicular at center node. Variation of gauge thickness and loading characteristics. Mass density 7.8E-9 Ns^2/mm^4 (curve 1 = fast loading)

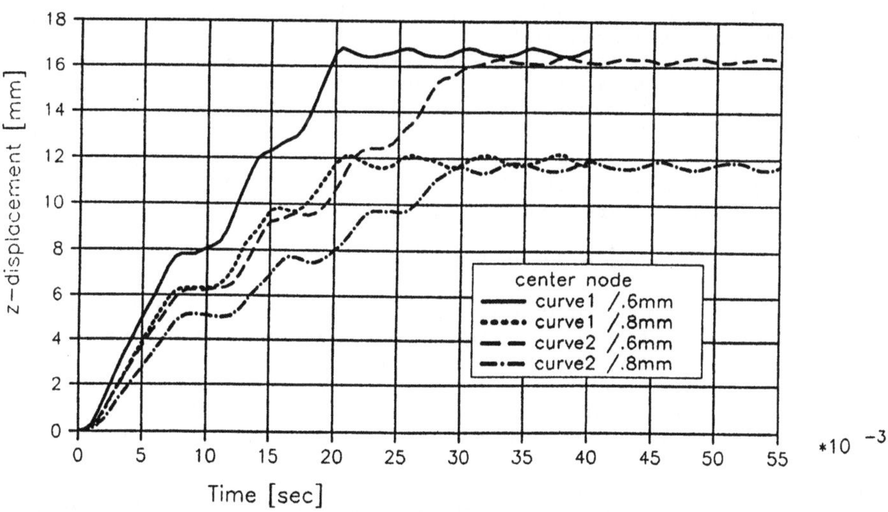

Figure 3: Square plate loaded perpendicular at center node. Variation of gauge thickness and loading characteristics. Mass density 7.8E-8 Ns^2/mm^4 (curve 1 = fast loading)

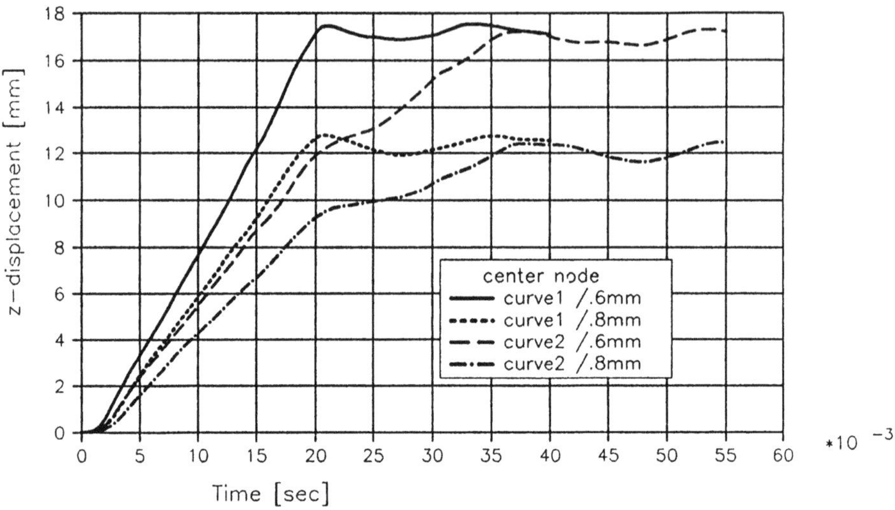

Figure 4: Square plate loaded perpendicular at center node.
Variation of gauge thickness and loading characteristics.
Mass density 7.8E-7 Ns^2/mm^4 (curve 1 = fast loading)

This maximum value remains constant after 20 and 30 ms. In the case of **0.6 mm** plate thickness, it can be seen that using the 10-times increased mass density the final deformation increased by **4%** (from 16 mm deformation for the initial mass density value to 16.6 mm) and using the 100-times increased mass density the difference became **8%** (final deformation is about 17.3 mm).
The difference becomes smaller if the slower loading (curve 2) is used. It was about **2.5%** with **10 x** ρ_0 (from 11.6 mm to final deformation of 11.9 mm) and about **6%** with **100 x** ρ_0 (12.3mm final deformation).
The natural frequency of the system is influenced of course and becomes evident by the different oscillation frequencies. The effect of the different loading slopes is not clearly visible in these diagrams. But this can be clarified if the whole system is reduced to a simple one-mass-spring system which is loaded by different time dependent linear force functions f(t),

$$m\ddot{x} + kx = f(t) .$$

This reduced system provides some interesting results. These results can be found in **figure 5** showing the solution of such a system with the different slopes of the linear force function f(t) without changing mass and stiffness values of the system. The reduction of the loading slope leads to significant amplitude reductions of the oscillation around the static solution. It is evident that the use of an infinitely small slope of the loading would lead to the exact static solution without any oscillations. This would allow the mass variation (increase) without influencing the final deformation.

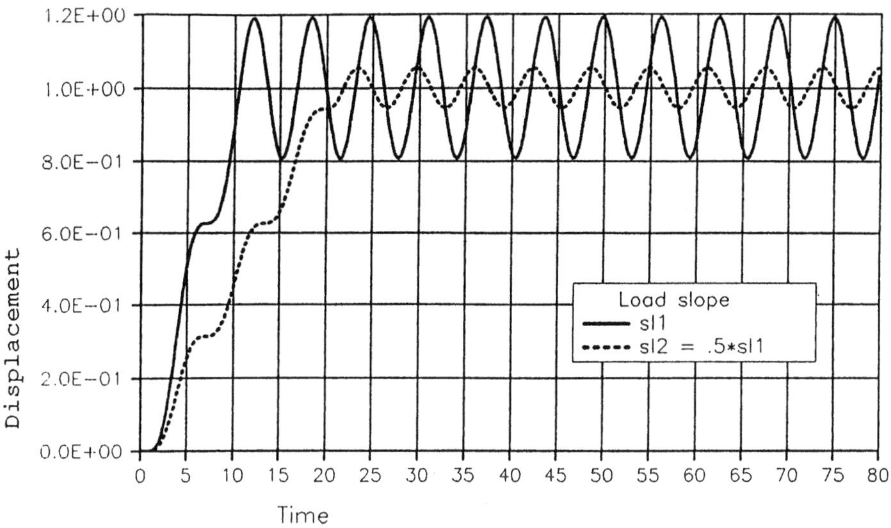

Figure 5: One – Mass – Spring – System
Linear load increase to 1. – constant load afterwards
Different load slopes – sl1 sl2

In a production run, however, a compromise has to be found between the calculation time and the accuracy of the results which may be influenced by the introduced inertia effects. In other words, some inertia effects have to be accepted in order to keep computation time small.
Deformed shapes of the regarded plate analyses are in this well defined case very similar for all analyses.

Analysis of the seat belt anchorage procedure.

Model description and analysis.

The seat belt anchorage location is a body structure part which has to withstand a very high loading in a quasistatic test procedure. Analytical methods have been proved suitable to reduce the costly and time consuming testing.
The finite element model used for the explicit analysis presented here, is a c-pillar assembly model shown in **figure 6a**.
For the element size of this model the stable integration time step for the explicit time integration scheme has been taken into account. The stable time increment depends on the sound speed in the material, i.e. on the material mass density and on the finite element size: the smaller the element the smaller the time increment has to be chosen for a stable solution. The smallest element size length in this model has been modelled larger than **3mm**.
The anchorage point, i.e. the loading point of the structure has been modelled to be the seat belt loop position and has been connected to the surrounding parts of the structure with **24** elastic bar elements. The mass density of these bar elements has been reduced by one order of magnitude in order to reduce inertia effects of these stiffeners.

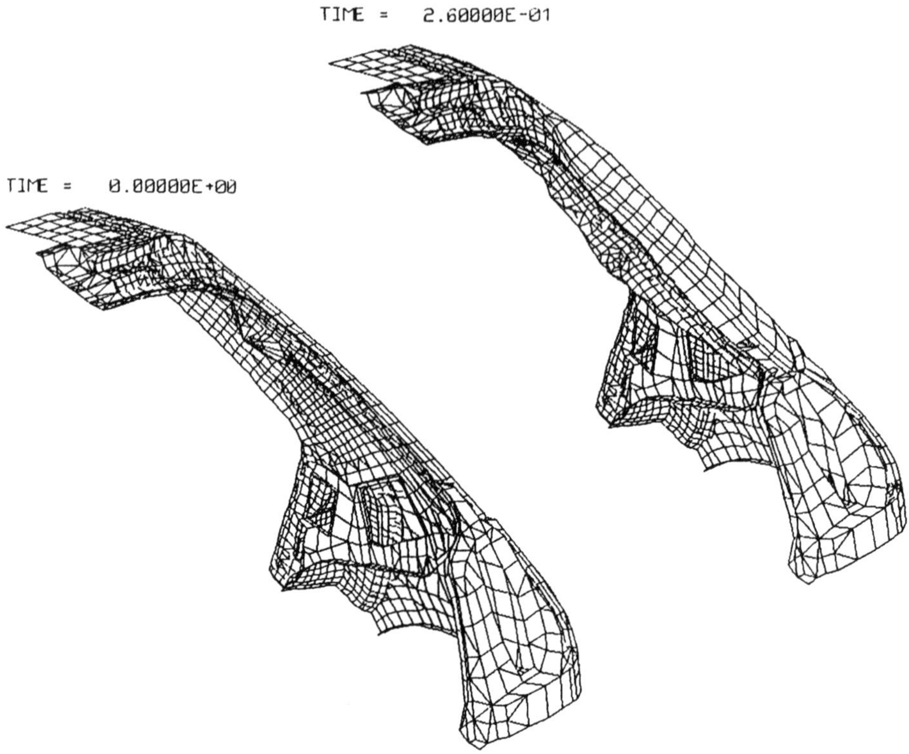

Figure 6a,6b: Undeformed and deformed c-pillar assembly

The influence of the structure loading via a time dependent load curve has been discussed earlier. Having in mind the discussion the loading in this FE-analysis has been introduced by the load-time curve shown in **figure 7.**

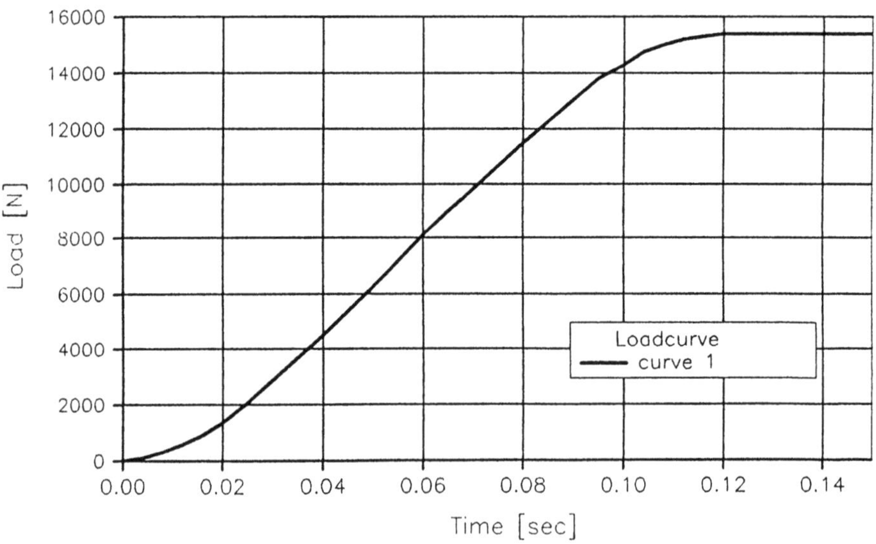

Figure 7 : Load – time curve for the load applied in the seat belt anchorage analysis for a c–pillar assembly.

This selected curve provides - as a compromise - a good solution in terms of time consumption and acceleration pulse.
The orientation of the load can be defined by using the load curve together with specified scaling factors for each direction x, y and z [2].
Pinned boundary conditions (fixed translational degrees of freedom, released rotational degrees of freedom) have been used at the cut sides of the assembly, which are at the roof-panel, at the rocker, at the rear-panel and rear wheel housing.

Self-contact area has been defined in a box around the seat belt anchorage point in order to avoid sheet overlapping in highly deformed areas.

The yield stress of the material used is **252 N/mm^2** and the stress strain curve is shown in **figure 8.**

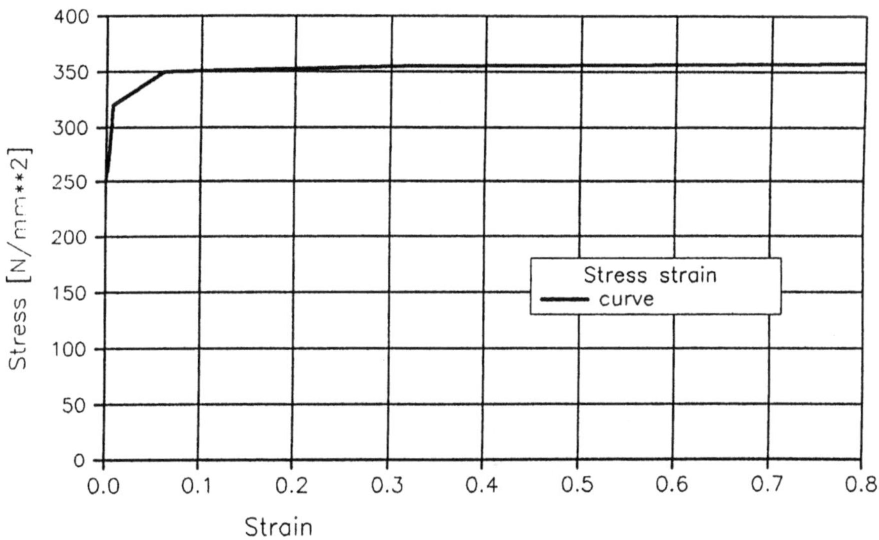

Figure 8: Stress strain curve used for the material in the seat belt anchorage analysis.

This curve represents a measured stress strain curve from the concurrently tested c-pillar parts [2]. It is worth mentioning that hardening above **350 N/mm^2** is very low in this material.
Three different mass density values have also been used for the investigation. Namely ρ_0 **= 7.8 * 10^{-9} Nsec^2/mm^4** which is the mass density value for steel, **10 *** ρ_0 and **100 *** ρ_0.

Results of the seat belt anchorage analysis.

It has been mentioned earlier that the oscillation along a static solution can be regarded as a measure of the dynamic effects in the system. If the inertia effects become small the oscillations diminish.

The introduced dynamic effects due to the loading are in this treated seat belt anchorage analysis very small and do not influence the results significantly.
This fact can be seen in **figures 9, 10 and 11** which show displacements of the the load attachment point in x, y and z directions respectively.

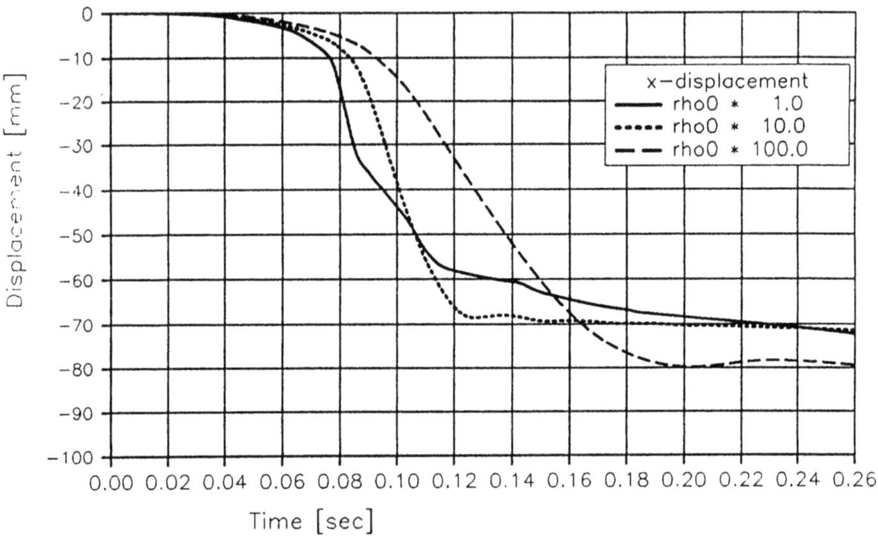

Figure 9: Seat belt anchorage analysis for c-pillar assembly
Displacement of anchorage point displacement
with different mass density values.

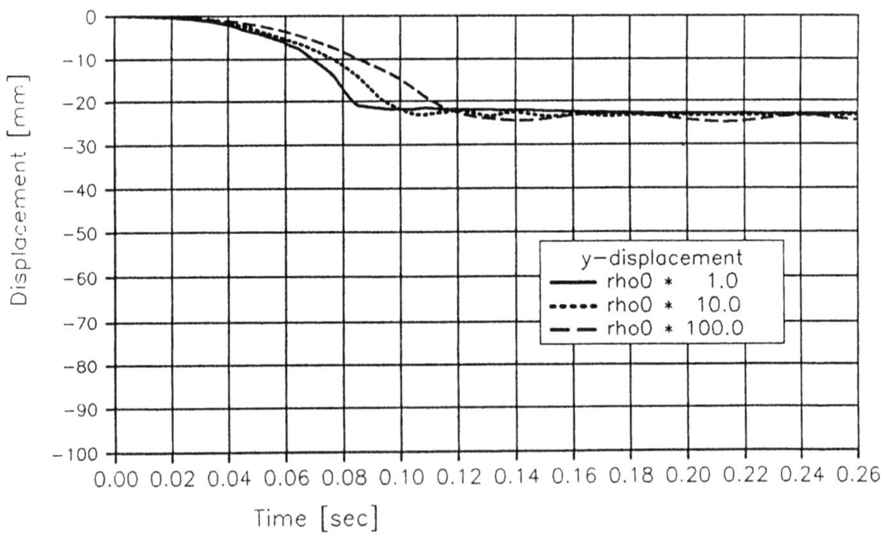

Figure 10: Seat belt anchorage analysis for c-pillar assembly
Displacement of anchorage point displacement
with different mass density values.

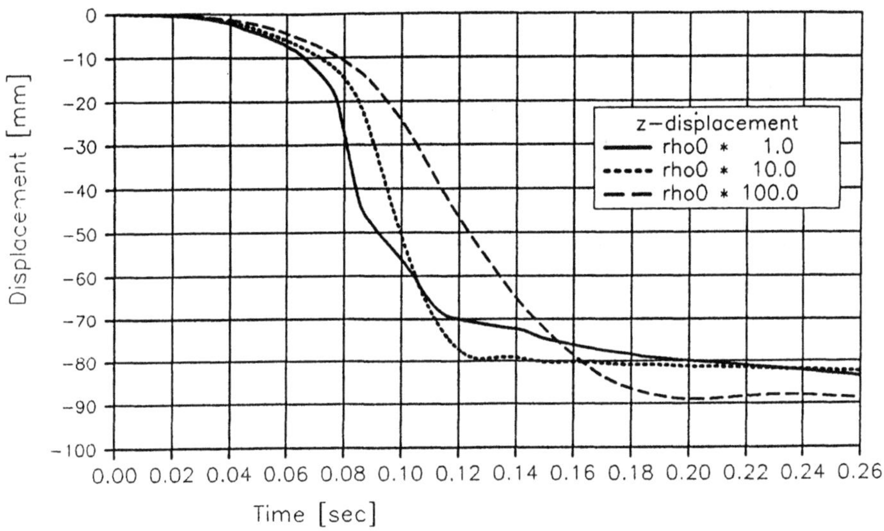

Figure 11: Seat belt anchorage analysis for c-pillar assembly
Displacement of anchorage point displacement
with different mass density values.

In this particular case it is not even necessary to damp out the oscillations with an available special option in the crash code [3].
The use of the damping option, in fact, is difficult for this case since the oscillation frequency is not clearly visible to serve for the identification and calculation of critical damping values.

The most important observations which need to be discussed here are firstly the continuous movement of the load attachment point beyond the time where the load has reached the maximum value and has been kept constant after it (120 msec) and secondly the difference in the maximum displacement of the load attachment point due to the different mass density values.

The simple geometry of the preliminary study with its well defined boundary conditions yielded to a constant remaining deformation level after the load was kept constant. This was not observed in the c-pillar analysis. The anchorage point in this analysis keeps moving in the main pull direction which is the x- and z-direction, respectively. Especially for the analysis with the initial mass density value this movement is very significant.
An explanation for this fact can be given regarding the stiffness of the structure.

Depending on the material behaviour, i.e. the stress strain curve description, the deformation pattern in the deformed structure can be influenced. Different plastified zones in combined elastic and elastic-plastic buckling processes can

lead to different paths at bifurcation points.
(These bifurcation points are anyhow more or less met by chance in an explicit integration method [2].)
One path could be a path with decreasing load deflection characteristics which might be met if a continuous cross-section collapse causes a continuous stiffness change. The other path could be a path with an equilibrium to the applied load.
In our case of the c-pillar assembly the non-constant displacement level after 120msec loading is an indication of such a structure stiffness change in the c-pillar. Referring back to the one-mass-spring system this behaviour can be simulated by decreasing linearly the stiffness value **(figure 12)**.

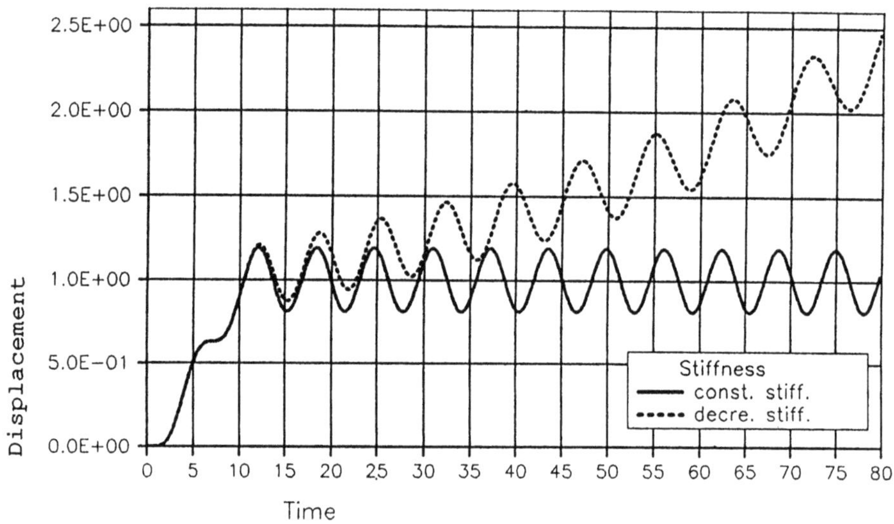

Figure 12: One – Mass – Spring – System
Linear load increase to 1. – constant load afterwards
Constant stiffness up to 10 sec – linear decrease afterw.

The same behaviour as in the seat belt anchorage analysis can be observed where the attachment point continues to move.
The slope of this continuous anchorage point deflection becomes smaller if the mass density value becomes larger. The higher mass has a stabilizing effect.

In a concurrently performed test the anchorage point displacement remained constant with constant load application. An explanation for the differing analytical results is the fact that the material behaviour and gauge thickness in a fabricated part is not exactly reported and known. Structure parts and special highly deformed areas resulting from the deep drawing process get different gauge thicknesses and sometimes different material properties. Since not known, these different material parameters are not taken into account.
In has been also found later in other analyses that other

material behaviour - having a larger continuous increasing hardening for instance than in the material used - led to constant remaining deflection after the load has been kept constant.

In the test the measured z-displacement of the anchorage point is about **60 mm** in the negative z-direction. In the analysis this figure is not constant as mentioned. But having in mind the discussion above and extrapolating the results with these parameters (especially the material property) the deflection could be assumed to be somewhat between 70 and 80 mm. Right after 120 msec the difference between the deflections with ρ_0 and 10 x ρ_0 is about 10 mm, which is about **14%** more deflection with the 10times increased mass value.
This figure becomes about **25%** if the 100times mass value is taken for a similar comparison.

For the particular problem the analysis has been continued up to 260 msec in order to assess the continuation of the anchorage deflection. It is not always necessary to perform the analyses that far. In the presented investigation the CPU-times for the three different analyses can be found in **table 1**. Even the run with the initial mass density appears to compete with comparable implicit time integration analyses in terms of CPU-time consumption.

Table 1: CPU-time requirement in [sec] for analyses up to 260 msec.

Analysis	**Mass Density**	**CPU - Time [sec]**
1	ρ_0	**61750**
2	**10 *** ρ_0	**19660**
3	**100 *** ρ_0	**6293**

Conclusions & Recommendations

It has been shown that quasistatic solutions can be approximated with an explicit time integration programme if appropriate loading conditions and a moderate mass density increase have been selected. A correct balance between loading characteristics and mass density increase serves to minimize the dynamic effects. If the loading is smooth the mass density value can be regarded as a "free" parameter and can be increased to a certain amount without influencing results too much. This enables reduction of computation time by increasing the stable integration time increment.
It has been shown that the CPU-time can be reduced considerably (**>30%**) without deviating too far from the solution without mass density increase, if an analysis with 10times mass density increase is being performed. The difference between test and an analysis without any mass density change is small in an engineering sense. The deformations for this type of analyses are always larger if the mass is being increased and can be regarded therefore as conservative figures.

Since dynamic effects can not be avoided completely material behaviour and loading conditions are crucial. The material properties are important for the buckling behaviour and need to be accurate and in line with the real material properties in the structure. The loading characteristics need to be analysed more intensively because the introduced dynamic effects into the system could lead to non-realistic displacement figures in conjunction with the mass density increase. Both parameters, material and loading, influence the solution path at possible bifurcation points.

Although some more investigations need to be performed concentrating on the above mentioned items, it is shown, that this method is useful to approach quasistatic solutions for a fast preliminary investigation of design alternatives.

Referencies

[1] Proceedings to the VDI conference : "FE-simulation of 3-D sheet metal forming processes in automotive industry", Zürich, May 1991.

[2] fka, F Herrmann: "Berechnung des Festigkeits- und Verformungsverhaltens einer Sicherheitsgurtverankerung mit nichtlinearen statischen Finite - Elemente -Methoden", June 1990.

[3] PAMCRASH manual, ESI GmbH Germany.

[4] Dr. B. Kröplin, M. Wilhelm, M. Herrmann : "Unstable phenomena in sheet metal forming processes and their simulation ", VDI conference, Zürich, May 1991.

Simulation-Assisted Method for Evaluation of Engineering Design Criteria for Car Structures and Components

Mr. Stig Wester

Volvo Car Corporation

ABSTRACT

This paper describes recent work at Volvo Car Corporation to make a systematic approach to the definition of load cases for cars. The intention of the project is to find critical driving situations and to analyze these situations with the assistance of simulation models and suggest tools for analysis in the Engineering Design Process. For every type of situation and for each level of design refinement there should be a range of suitable models. In early stages of the design process the models may be crude without too much details and in the final stages the models are more complex up to the inclusion of FEM models. For instance an early model of the suspension shouldn't contain any geometric details but wheel movements could be described by curves implemented as polynomials, functions or tables in the model.

INTRODUCTION.

Because of the pressure from new regulations, market and competition, car manufacturers have to produce new and more sofisticated models in shorter time. Traditionally the developement has heavily relied on testing procedures but now there is a change towards the usage of analytical methods. One other reason for this new trend is the development of better computerized tools. Anyhow I don't think there will be less testing but the contents will change.

One type of analytical tools is simulation, which is used for dynamic problems when it isn't possible to derive an explicit formula for the design parameters.

The usage of analytical tools varies from area to area within the company depending on test and personal resources.

The current project described in this paper has the objective to collect available methods, produce new methods and document them in a systematic way for the prediction of forces in some significant driving situations.
As the project has just started the paper only contains some basic ideas and an example for driving over a potthole with different velocities. Results from a simple simulation model is compared with test results.

LOAD CASES.

The critical load cases are defined by strategic decisions, by legal requirements or regulation proposals. Some load cases are straight forward but others are more general and require further exploration. In this more general case one could search for a point in the parameter space that gives highest requirement for some part and define this as a critical load case.

So far we have suggested the following subdivison of load cases :

+ Road generated forces:

= Pothole.
= Speedbump.

+ Manouver forces :

= Sudden clutch pedal slip off.
= Braking in pothole.
= Slip against pavement.

On the test track a lot of obstacles have been constructed to simulate theese situations and the car is running over them following a special program for each type of car.
The first object of the analytical approach is to predict the behaviour of the car on theese test track obstacles.

MODELS.

One could consider three levels of models from concept model without geometric details to an intermediate level with mechanisms included for load distribution up to detailed models even with FEM parts included.

Concept model.

With the concept model it is easy to do parameter studies. The answers from calculations come quickly and the behaviour could bee easier to survey. This is specially important if the model due to nonlinearities behavies in a complicated way. The results could bee used for calculation of forces at fastening points between chassi an car body. The load distribution in the car body could then also bee calculated with a beam model. The simulation tool for these models is CSSL-IV due to the high level of interactivity in its "RUN TIME ENVIRONMENT" where you run experiments with the model.

Intermediate Model.

This model could be a model of the front suspension system with relavant components included to calculate the behavior and to get the forces in each component. The simulation tool for the Intermediate Model is ADAMS because of the possibility to model mechanisms. These types of models take longer computation time because of their high complexity.

Detailed model.

In this model the chassis systems are modelled and the car body could bee a FEM model with several eigenmodes included. This model will take long time to run but colud be used for a final check of the behavior of the system before starting tests or to check that the approximations made in the simplified models are valid.

Example CONCEPT MODEL FOR POTHOLE.

As an example of an concept model I would like to present a model for the calculation of forces when driving over a pothole. The model for the car body is a rigid body with 3 dof (vert, roll, pitch). Each wheel is modelled by one mass with 2 dof (vert, longitudinal). The inclusion of roll and pitch is because only one side of the car hits the pothole.

Vehicle model :

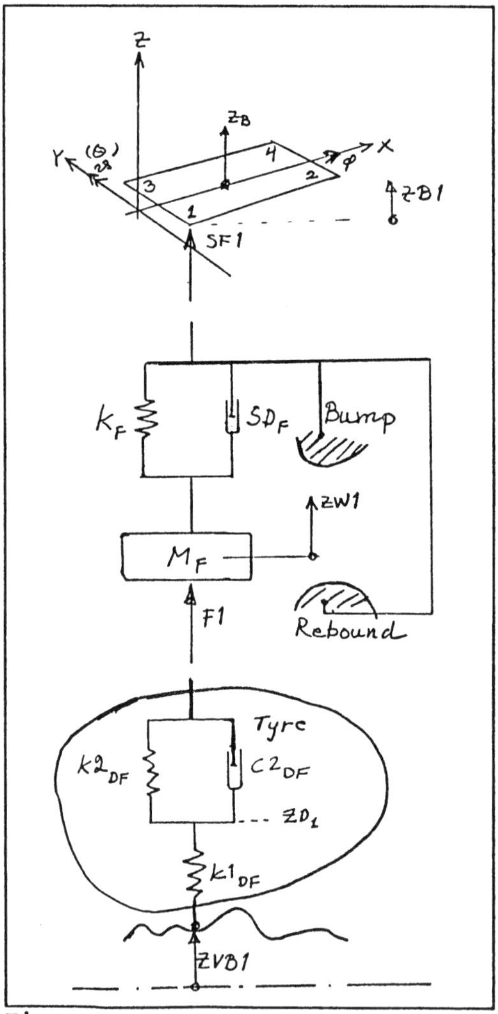

Figur 1

Z_B	Z-coordinates for CG.
ϑ	Pitch angle.
φ	Roll angle.
ZB_i	Z-coordinates for corners.
$BMPK_i$	Bump forces.
RBK_i	Rebound forces.
$DMPK_i$	Damping forces.
FJK_i	Spring forces.
SF_i	Sum of forces acting on car body.
FH_i	Spring compression.
ZW_i	Z-coordinates for wheel.
DZW_i	Z-vel. for wheel.
$DDZWi_i$	Z-acc. for wheel.
MF(MB)	Unsprung mass front(rear).

Index i :

i	Position.
1......	Front left.
2......	Rear left.
3......	Front right.
4......	Rear right.

Vertical tyre model :

The vertical tyre model is a point contact against a virtual road profile. The virtual road profile describes the movement of the center of the wheel when it moves quasistatically over the obstacle with nominal vertical force at the wheel center. The actual dynamic deflection is then calculated from the dynamic forces with the linear spring model you see in fig.2. The assumption is that the virtual road profile won't change significantly with vertical load and that theese minor changes to the curvature won't have much influence on the calculated results. The contact force is made > 0 so the model may loose contact with the virtual road profile.

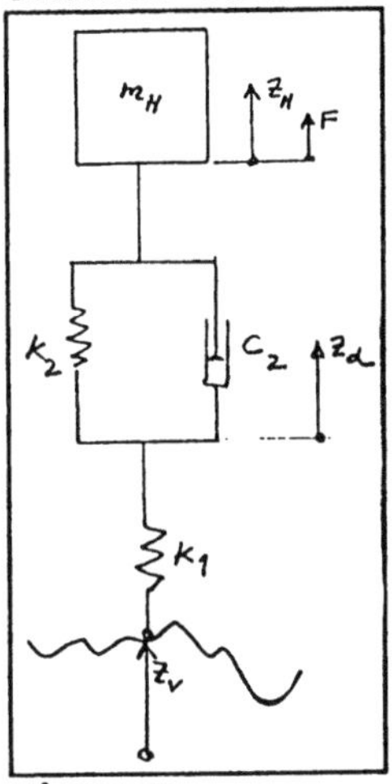

Figur 2

$$F=k_1\times(z_v-z_d) \tag{1}$$

$$F=k_2\times(z_d-z_H)+c_2\times(\dot{z}_d-\dot{z}_H) \tag{2}$$

Elimination of F from equation (1) and (2) gives :

$$\dot{z}_d=\dot{z}_H+\frac{k_1\times(z_v-z_d)-k_2\times(z_d-z_H)}{C_2}$$

Reasonable values for the constants for standard tyres could be :

k_1 = 234 000 N/m , k_2 = 780 000 N/m and
c_2 = 125 000 Ns/m.

The frequency response for the tyre model is :

$$\frac{F}{\delta}=\frac{k_1\times k_2}{k_1+k_2}\times\frac{1+j\times\frac{\omega}{\omega_1}}{1+j\times\frac{\omega}{\omega_2}}$$

where $$\omega_1=\frac{k_2}{c_2}$$

and $$\omega_2=\frac{k_1+k_2}{c_2}\ .$$

This tyre characteristic has a low frequency stiffness of $k_1\cdot k_2/(k_1+k_2)$ and a high frequency stiffness k_1. Standard values above give a low frequency stiffness of 180000 N/m and a transition frequency of 1 Hz. With an unsprung mass of 40 kg one gets an eigenfrequency of 12 Hz for the unsprung system .

This tyre model has the advantage that it only needs the vertical road profile as input and not its derivative.

Horisontal tyre model :

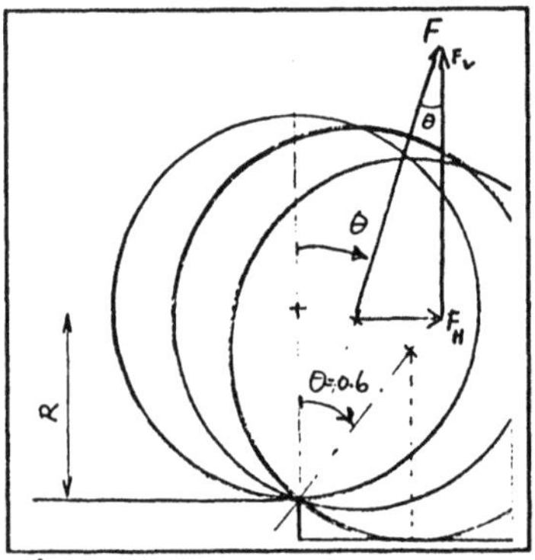

Figur 3

The assumption of the tyre rolling as rigid ring over the edge of the pothole gives the following formula for the relation between rolled distance and rotation angle : $\sin\theta = \frac{s}{R}$

With the assumption of no braking the resultant force goes through the center of the wheel. This gives :

$$\frac{F_H}{F_V} = tg(\theta) = \frac{\frac{s}{R}}{\sqrt{1-(\frac{s}{R})^2}}$$

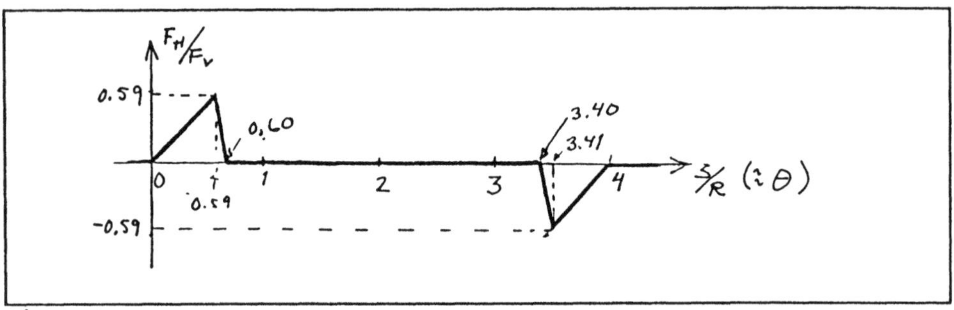

Figur 4

The function $F_H / F_V = g(s / R)$ is entered through a CSSL table or is calculated according to the formula given above by a FORTRAN subroutine .

Model for horisontal wheel movements :

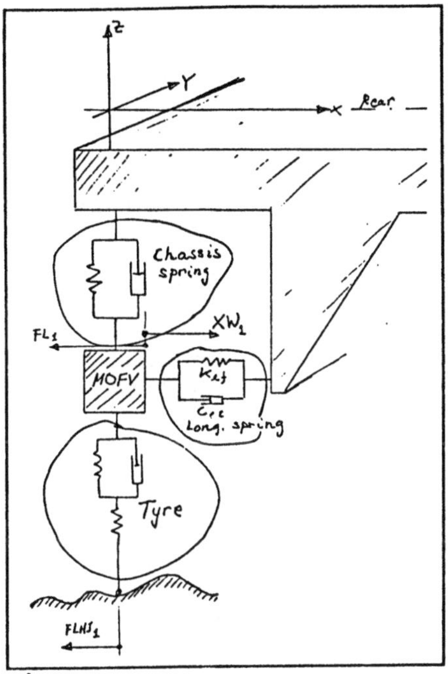

Figur 5

Force directions and coordinate systems are defined in the adjoining figure.

Input data for the horisontal spring is eigenfrequency (f_l) and relative damping (ζ). From these constants one can calculate spring constant and damping constant.

$$k = m \times \omega_0^2 = m \times (2 \times \pi \times f_l)^2$$

$$C = 2 \times \zeta \times \omega_0 \times m = 2 \times \zeta \times 2 \times \pi \times f_l \times m$$

The vertical forces generated by the horisontal forces (antilift etc) are taken care of by a proportionality constant.

RESULTS FROM TEST AND CALCULATION.

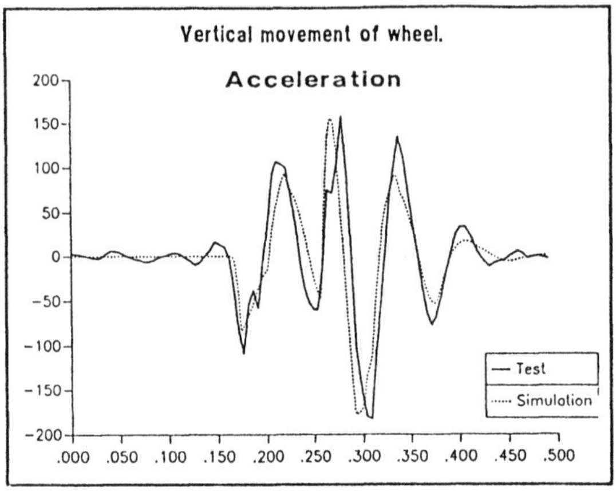

Behaviour of cars on our test tracks have been measured very caryfully. I took one of these results for a first validation of the concept modfel. Because fairly narrow classification of parts it was possible to just use specifications from drawings as input data to the model. The report from the test contains information on which class each component comes from.

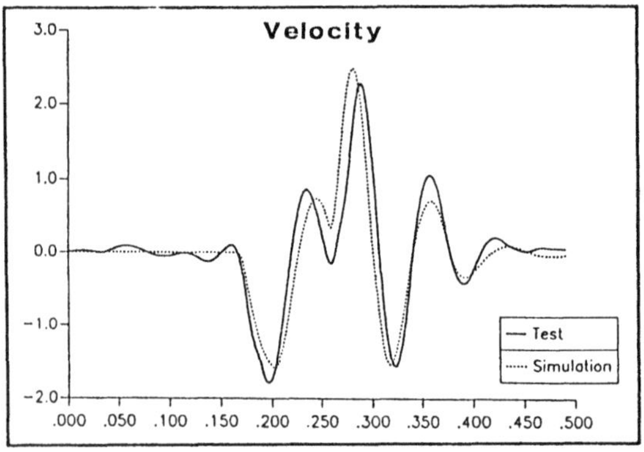

To make a more careful validation it is necessary to measure the components mounted in the test car and to measure some more variables in the test run. As can be seen from the graphs there are some differences but in principal the test and calculated results agree quite well.

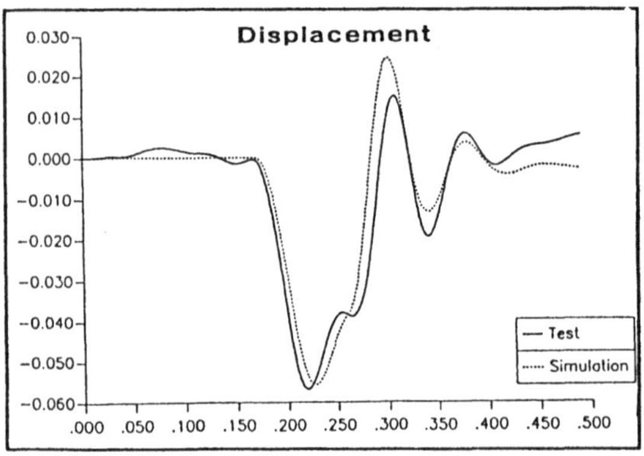

The measurement of vertical wheel movement contains only vertical acceleration. The velocity has been recived by integrating this signal.

$$v(t)=\int_0^t (a(x)-a_0)\,dx$$

a_0 was necessary for correction of a small zero drift in the measurement equipment.

Integration of the velocity gave the displacement.

$$s(t)=\int_0^t v(t)\,dt$$

As seen in the graph below there is a larger difference between test result and simulation result for the longitudinal force. There is anyhow a large difference between different runs over the same obstacle and between runs over different but nominally equal obstacles on the test track so this requires furher investigations.

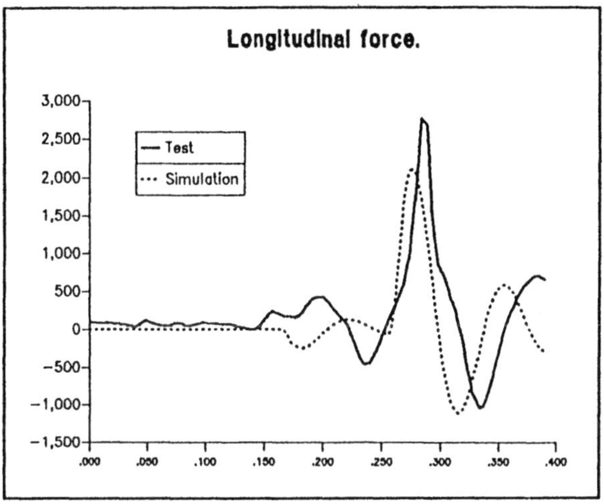

CONCLUSIONS.

The results so far look promising. To get better results from the concept model there still remains a lot to do. Anyhow nothing up to now contradicts the usefullness of the suggested stepwise refinement process. One usage of these models is also to check the testing procedures.

This type of stepwise refinement modelling raises the requirement of a block orientation of the simulation language. That makes it easier to change one part of a model to a more detailed one without any possibility of collisions in variable names. Further on it would be nice to have a possibility of easily switching one part of a model at run time. The CSSL language I have used doesn't support this very well. Anyhow I like the flexibilty of the CSSL environment and that it has FORTRAN as a subset and that it is easy to include small FORTRAN subroutines by inserting them at the end of the input deck.

Engine

CAD/CAE/CAM – Integrated Applications in Engine Development

Dr.-Ing. Gotthard Rainer, Head of "Computational Structural Analysis",
AVL - Company for Internal Combustion Engines and Instrumentation Ltd.,
Graz, Austria

1. Introduction

The necessity to achieve ever shorter development cycles demands more and more the use of "simultaneous engineering" processes. To link together the phases of design, simulation and machining (prototyping) increasingly requires extensive graphical tools, the combination of various calculational and experimental methods and advanced production techniques. Furthermore a high degree of integration is demanded.

These are basic requirements for the development of high-tech products within extremely shortened development time. Therefore it is necessary to develop methods which allow optimizations as early as at the design stage to achieve an "analytical prototyping" by simulation of functionality and working conditions of the product.

AVL have following product branches:

- Engine development
- Development of equipment for
  - Engine Instrumentation
  - Ballistic Instrumentation
  - Medical Instrumentation.

AVL aim to be active in a market segment only in case

- **high-tech products** are in demand and it is certain that owing to AVL-expertise
- a **product of excellent quality** can be offered.

To achieve this goal both highly qualified staff and sophisticated tools are necessary. These tools consist of hardware and software components which are only partially available on the market or have to be adapted to the specific needs of AVL. This results in the following requirements of various specialized departments:

- Search of the best tools on the "free market" and adaptation to the specific needs of AVL
- Supply tools which base on AVL expertise and cannot be bought externally
- Optimal application of the tools to the respective tasks

2. CAD / CAE / CAM

The "CA*" technologies are an important part of the new engineering techniques. The success depends to a high degree on the application of CAD. CAD as an aid for the design department is economically not viable on its own. Its actual merits are rather in the connection with calculation and manufacturing. Therefore it is necessary to evaluate a CAD-system covering the needs of integration and optimal adaptation for special applications.

AVL's most important motivating forces to apply CAD were:

- to improve the quality of the engineering services
- to get a better integration of all stages of development
- to shorten the "time to market"

2.1 Requirements of CAD

CAD-application to engine development needs extensive abilities as regards software and hardware, owing to the complex geometries of each component.

Since AVL use CAD not only as a kind of "drawing board substitution" but also for the design, it was absolutely necessary that the CAD-system to be installed not only has a 3D-wireframe model but also an efficient surface and solid model [1].

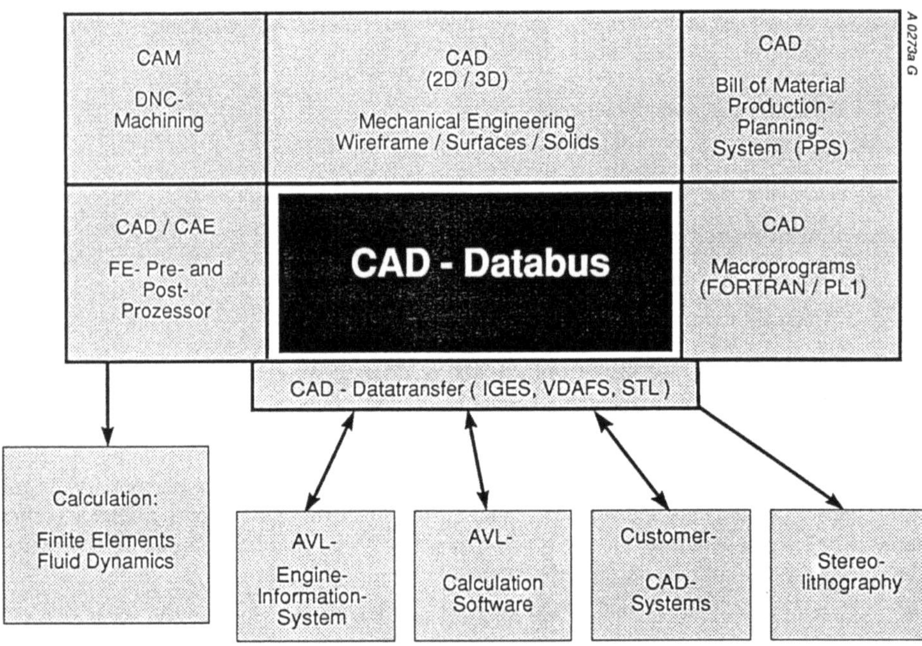

Fig. 1 Integration of CAD into AVL-Engineering Software Environments

Moreover, AVL carry out engine development for a large number of customers all over the world who require CAD data to be transferred to their own CAD systems. Because these customers use different systems, standardized interfaces (IGES, VDAFS) must be available for the data transfer.
For the in-house data flow it is necessary to provide complex calculation methods and CAD-geometry data for the manufacturing.
It is also important that macro programs can be created and used.

Fig.1 shows the necessary integration of CAD into AVL-Engineering-Software environments for the support of the product development.

AVL are using the CAD-System BRAVO 3 (Schlumberger).

2.2 Steps to increase the efficiency of a CAD system

The CAD systems usually available on the market are "General Purpose" systems. They may be applicable in many areas, however, in defined product environments they can only be applied optimally if specific adaptations and additional extensions are made.

According to application at AVL following steps have to be considered:

- **Adaptation of CAD-system parameter:**
  As an example the definition of standards is mentioned, e.g. DIN or ISO.
- **Assembly of often used command sequences:**
  Often used command sequences are to be assembled to simple commands and made available to the user.
- **Graphic procedures for similar geometries:**
  Many CAD-systems offer a "Graphic Language", with the help of which the user can write menu-controlled programs for the creation of geometries. These languages are easy to use but very CPU-time consuming because they work interpretatively.
- **Building up of geometry libraries:**
  During the development of new products there are few "novel designs" but more "design modifications". The novelty most often consists of modifications of details of engine components and, therefore it is useful to store the completed projects in the form of "geometry libraries".
  The structure of geometry libraries requires a systematization of designs. For an efficient application it is necessary to have a special data base in the CAD-system (drawings' administration).
- **Building up of macro programs:**
  The creation of macro programs is of all other possibilities the best way to define component geometries. They are not only applicable in the design department but also for the creation of FE-structures, [2,3]. The connection between macro programs and CAD-system can either be established by subprograms, which are made available by the CAD-system for data storage and graphic representation, or by file-oriented data transfer.

3. Integration of CAD and Finite Element Calculation

At AVL FE-Calculations have already been in use for many years [4 to 8] and are State of the Art in engine development.

To achieve more and more complex development goals and to shorten the development time it is necessary to make efforts in all stages of FE-calculation which are

- Preprocessing
- Calculation
- Postprocessing

3.1 Preprocessing

Fig. 2 shows various possibilities to connect CAD and FE-Calculation. The ideal situation is to have an integrated system for both applications. Therefore AVL decided to install a FE-Preprocessor which is using the same database as the CAD-System. An additional advantage is the availability of enhanced CAD-Capabilities for building up complex 3D-Geometries. That point is very important with respect to the complicated geometries of engine components.

Disadvantageous may be that special FE-Preprocessors have more possibilities for manipulation of FE-meshes.

To cover the needs of engine calculation AVL has developed features including the definition of loading cases, boundary conditions, split lines, etc.

Preprocessing is mostly understood as creation of the FE-mesh. To perform complex FE-calculations FE-preprocessing is a combination of the following operations:

1) Abstraction of the original CAD-geometry for FE-mesh generation.
   (e.g. reducing solid geometry to Surface or Beam Elements, simplifying geometric shapes, etc.)
2) Definition of loading cases and boundary conditions.
3) Gathering of material properties.

For many cases in engine simulation points 1) and 2) are time defining points. For complex FE simulations it is economically not possible to automate these stages.

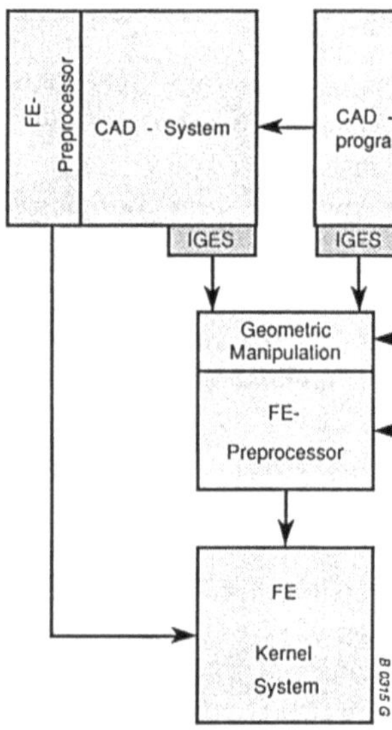

Fig. 2 Various Possibilities to Integrate CAD and FEM

But to bring FE-calculation closer to the design phase, the automatization of points 1) to 3) for "order one calculations" is absolutely necessary.

To meet this goal AVL works on a program to make available automatic FE-meshing including adaptiv remeshing and also design sensivity tools [9], **Fig. 3**.

AVL is using the FE-preprocessors GRAFEM and INTGEN.
GRAFEM is an integrated modul of the CAD-System BRAVO3.
INTGEN was developed at AVL.

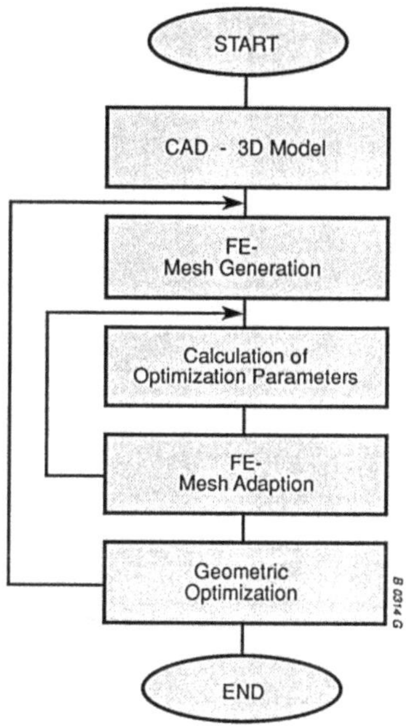

Fig. 3 Automatic FE-Optimization During Design Stage

3.2 Calculation

FE-Calculation tools are available as "general purpose" software on the market. Because of the improvements in the field of hardware and software, extensive calculations have been carried out during the last few years to cover the increasing demands in engine development. AVL has developed software, especially to optimize engine structures with respect to noise radiation (NVH). These calculation methods are very CPU-extensive. It was one of the reasons why AVL integrated a vector computer into the computer network as a number cruncher for technical computations, **Fig. 4**.

AVL are using the FE-systems MSC-NASTRAN and TPS10.

3.3 Postprocessing

The improvement of result presentation became very important parallel to the complexity of calculational methods. To evaluate results of high sophisticated calculations both powerful hardware and interactive graphical software are necessary. To interprete results of transient calculations also screen animation and video must be available.

AVL are using GRAFEM (Schlumberger), MOVIE.BYU and additional inhouse -programs for graphical presentations.

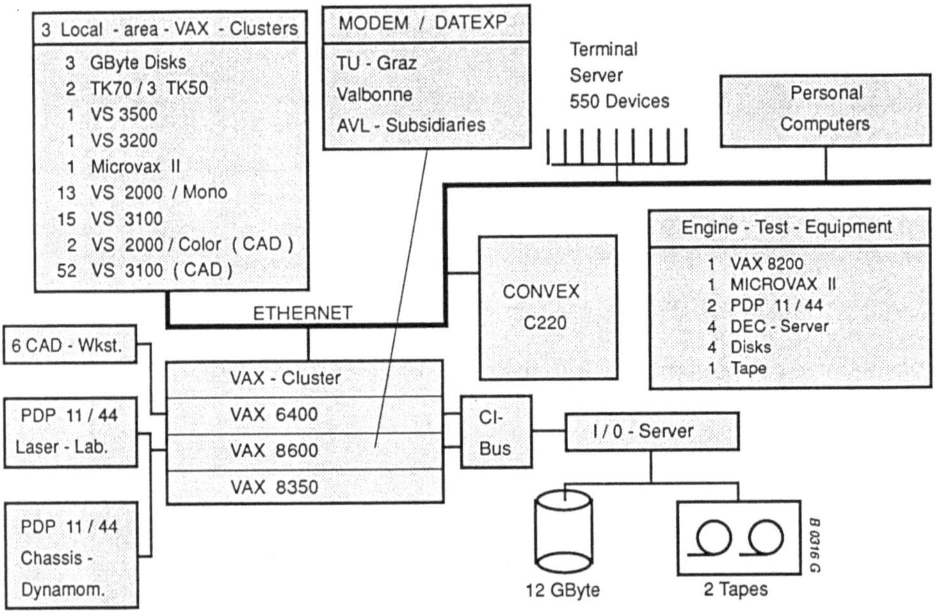

Fig. 4 AVL's Computer Network

3.4 Example: Integration of CAD and CAE: Dynamic - Acoustic FE-Calculation:

A characteristic application for complex FE-Calculations is the dynamic-acoustic simulation of engines [6,7,8]
The noise emission reduction has gained more and more importance in the last few years. AVL has developed a method which bases on the combination of FE-calculation, engine tests, acoustic measurements and result analysis, **Fig.5.**

Here each section of this procedure is described in brief:

- **Definition of the FE-model:**

  The FE-model contains two kinds of structural parts:

  - Structural parts which are directly taken from the design (e.g. cylinder block and cylinder head)

    By means of the CAD geometry provided by the design department the mesh which is necessary for the FE-calculation will be created. The FE-preprocessor has direct access to the geometry data so that there is no need for an additional interface.

  - Structural parts for which mechanical models are made (e.g. pistons, con-rods, crankshafts)

    To reduce the number of degrees of freedom it is necessary to describe the parts of the crank train by means of beams and bars. For this abstraction masses and mo-

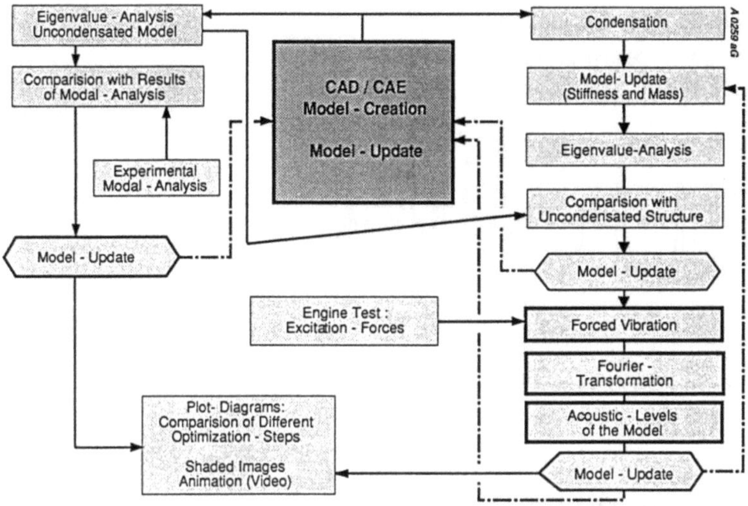

Fig. 5 Combination of Experimental and Calculational Methods to Optimize Engine Structures with Respect to NVH

ments of inertia as geometry data are needed. These values may also be obtained from the CAD system if a solid model is taken for the geometry.

- **Building up of the dynamic loads:**
  Owing to engine tests the pressure distribution in the combustion chamber have been made available. These values are taken over into the calculation either by a prototype of the engine to be optimized or -in the design stage- from test results of engines with similar specifications.

- **Carrying out the FE-calculation:**
  The FE-calculation of both the natural and forced vibrations are partially carried out by standard programs and partially by programs developed at AVL.
  Calculations are still done linearly, however, in the future also the nonlinear behaviour of the bearings (oil film) will be considered.
  The result verification of these calculations is carried out on the prototype with experimental results (modal analysis, acceleration measurements)

- **Acoustic analysis:**
  The programs for the acoustic analysis of the FE-results have been developed in-house and supply
  - speed and acceleration levels and
  - transfer mobilities

  The result verification also takes place with the help of experimental results.

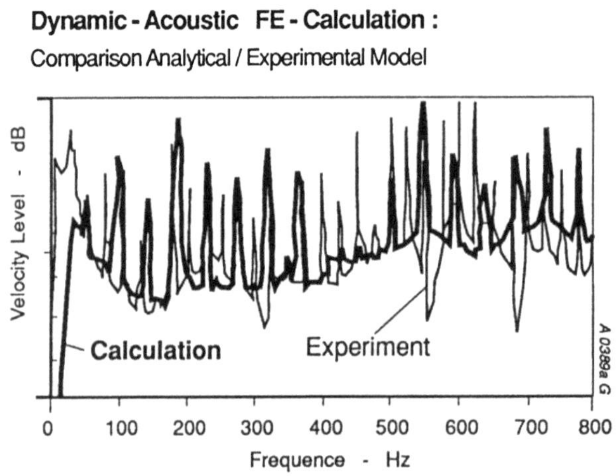

Fig. 6 Comparison of Experimental and Calculational Results

- **Result presentation:**
  Until recently the presentation of dynamic processes had been carried out by means of static plots. To make available visual aids for the quick and reliable result assessment the dynamic result presentation is used more and more.

  The most important ways of presentation are:

  - Diagram description for a direct comparison of results on individual points of various geometry variants, **Fig.6**
  - Coloured presentation to visualize the total result of a geometry variant, **Fig.7** and
  - Video recordings (animation) for the dynamic result presentation

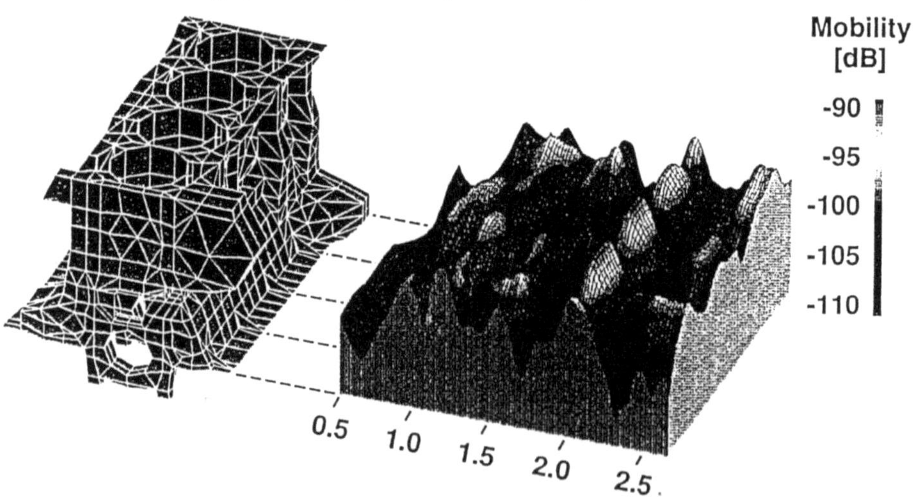

Fig. 7 Result Presentation to Review the Noise and Vibration Behaviour of a Crankcase

4. Example: Integration of CAD and CAM: Development and manufacturing of valve ports:

Compared with other engine components valve ports have special features:

While the customer usually gets calculation reports, drawings and test reports, as regards valve ports, he receives a core box and a port core in the form of "hardware".

The various steps during the development within the common procedure have been:

- Definition of the geometry according to prescribed boundary conditions and demanded specifications
- Creation of cross sectional drawings
- Manufacture of a core box
- Flow development and iterative optimization of the geometry

Fig.8 shows the whole conventional and future process of the development phases.

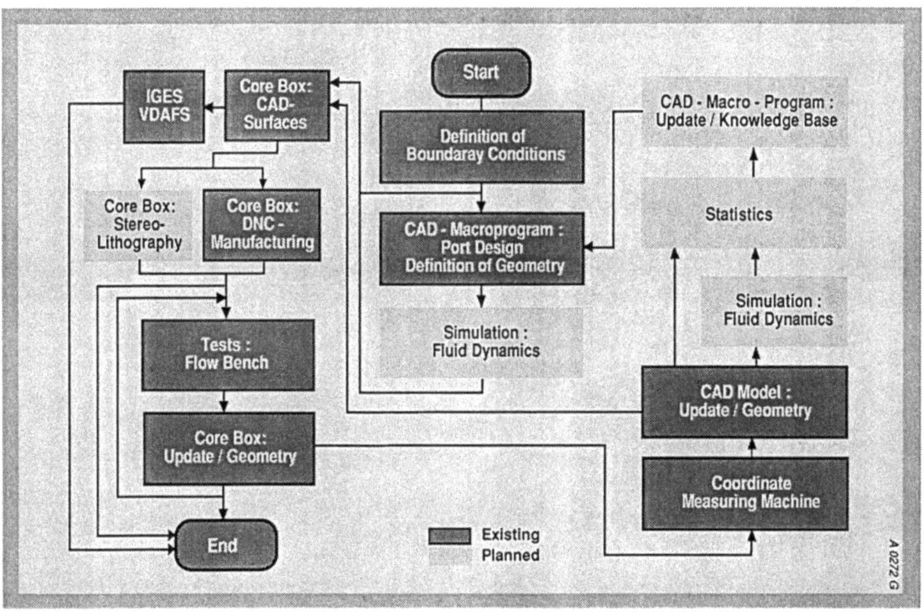

Fig. 8 Valve Port Development

The above mentioned development steps often were a "bottleneck" within the design phase.

Though the complexity of this engine component hardly allows the geometry description by means of macro programs, the integration of development and manufacture of valve ports was an important criterion during the CAD-system-evaluation-phase. It was thought to be entirely computer controlled and, thus be considerably time and money saving.

The aim of the use of CAD macro programs was to define the valve port in its geometry and to manufacture it on a milling machine.

In the first step the geometry of the valve port was calculated iteratively; here in this macro program the AVL-expertise necessary for the development of this component had been integrated as the "knowledge base". The calculated geometry is transferred to the CAD-system as a wireframe model in such a way that detail drawings can be made more easily, **Fig.9.** This step has already proved to be timesaving [10].

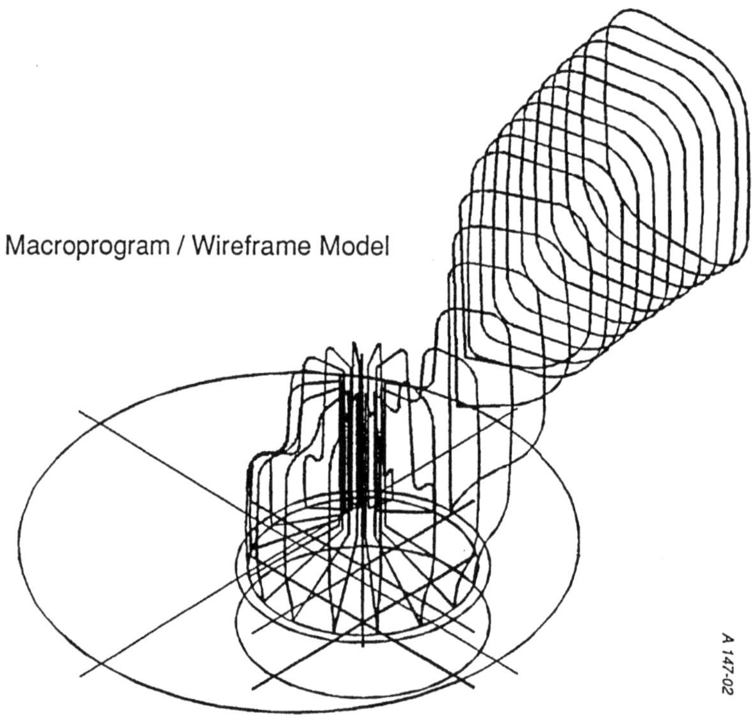

Fig. 9 Wireframe Model of a Helical Intake Port

As a second step procedures had been created in the CAD-system which enable a timesaving structure of a surface model from the wireframe model. It proved to be more efficient to manufacture the core box instead of the core since on the one hand, the manufacture is easier and, on the other hand the model has thus been available for the flow test, **Fig.10.**

During the flow tests geometry modifications have to be carried out. As a third step therefore

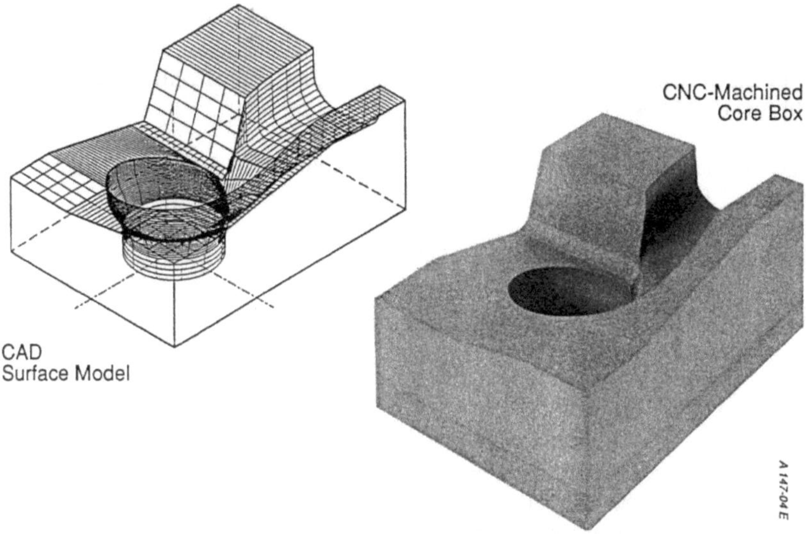

Fig. 10 NC-manufactured Core Box

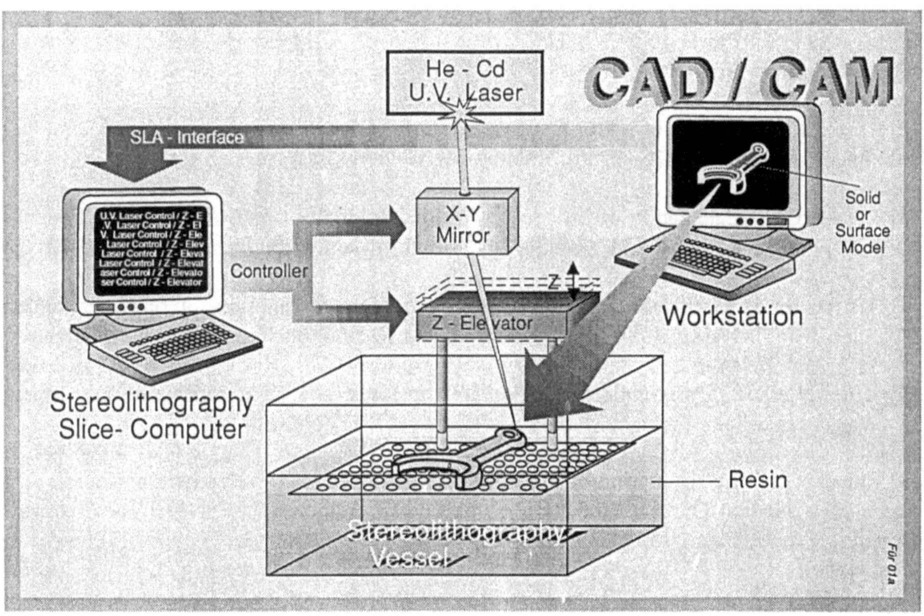

Fig. 11 Stereolithography

procedures have been installed so that the digitized modified geometry zones were integrated into the original CAD-surface model by means of a coordinate measuring machine. Instead of the hardware the customer can be supplied with a digital CAD-model which can be manufactured on his own facilities.

To reduce the production time for prototypes the stereolithography will be applied for a "Rapid Prototyping" [11], **Fig.11**. First tests have already been made.

Furthermore AVL use their own fluid dynamics programs [12], for the theoretical prediction of the flow behaviour. The preprocessing for the very complex calculations will also be integrated into the macro program.

5. Summary

The integrated methods based on "CA-techniques" for development and prototyping of engine components show that the combination of design methods and experimental and calculational systems are absolutely necessary for high quality products and shorter development cycles. However, for this purpose it can be seen that no efforts have to be spared in the fields of design, simulation and manufacturing for the "development of development tools".

6. References

[1] Rainer, G.: "CAD-Introduction and Application for the development of Engine Components", Congress Proceedings "CAD/CAM as a Part of CIM-Component", 1989, Vienna, ADV-Meeting

[2] Rainer, G.: "Structure of a Program System for the CAD-System independent Macro Development" Konstruktion 38 (1986), Volume 7 to 10

[3] Rainer, G.: "Calculation of Stresses in Welded Joints by means of the Finite Element Method", Forschungsheft FKM, Volume 74, Frankfurt/Main, Maschinenbauverlag, 1979

[4] Affenzeller, J., Gschweitl, E.: "Contribution to the Design Optimization of Water Cooled Cylinder Heads", ASME, 1986, Energy-Sources and Technology Conference, New Orleans

[5] Affenzeller, J.: "Deformation Investigations on Reciprocating Engines with Dry Liners by means of the Finite Element Method" ,MTZ 39 (1978), Volume 3

[6] Rainer, G.: "The Integration of the FEM into the Product Development by means of a Calculation Method for the Prediction of the Dynamic-Acoustic Behaviour of Engines", Congress Proceedings "8th Meeting in Reutlingen - FE in Practice", 1985, T-Programm Reutlingen

[7] Affenzeller, J., Priebsch, H.H., Rainer, G.: "Influence of Attached Parts on the Dynamic Characteristics on Cylinder Blocks", MTZ-Magazine 45 (1984)

[8] Priebsch, H.H., Affenzeller, J., Kuipers, G.: "Structure Borne Noise Prediction Techniques", SAE Technical Paper Series 900019, International Congress and Esposition, 12990, Detroit

[9] Rainer, G.:"Engine Development: Integration of Calculational Methods in CAD-Macroprograms", Proceedings, CAT 87, Computer Aided Technologies in Manufacturing, Stuttgart, 2.-5. Juni, 1987.

[10] Türtscher, A.:"Computational Development and Manufacturing of Valve Ports", Dissertation, 1991, Graz (non published)

[11] N.N "Stereolithography: Limits and Possibilities of a new Technology", Technologiebericht, Spectra-Physics GmbH, Darmstadt, 1989

[12] Bachler, G., Brandstätter, W., Steffan, H., Wieser, K.:"3D-Simulation of Fluid Mechanics-, Fuel Injection-, and Internal Combustion Processes", Meeting "The Working Process of an Internal Combustion Engine", Oct. 1989, Institute of Internal Combustion Engines and Thermodynamics, Tech. University, Graz

Drive Line and Control Simulation

An Integrated Driveline Analysis Program to Support Truck Design Process

Ing. Almondo	Ing. Balbi	Ing. Fischer
IVECO FIAT TOO-SS	IVECO FIAT PTO-DL	SDS SIMULATION

1. FOREWORD

The increasing competition on the market requires a reduction of development cycles of a new product. This is achievable through a 'Total Power Train Simulation' and through 'Simultaneos Engineering'.

It is reckoned that if a product with a market life of 5 years goes on the market with a time delay of 5 percent, it will cause a global profit loss of 30 percent. In the same time a 50 percent increase of the development costs, will reduce the profit only by 5 percent.

Starting with this input from Mr. Norzi, Driveline Engineering Manager, a review to the computation procedures for the driveline design was carried out.

Among the others, a new software was developed and implemented, to study the impact of design changes to the behaviour of drivelines for the 'drop clutch' test. This application will be illustrated from 3 points of view:

- Hardware and Software
- Engineering Approach
- Application

2. SOFTWARE AND HARDWARE

2.1 SOFTWARE

The project on the driveline simulation was started with a conventional approach, i. e. trying to use vendor provided general purpose simulation pakages, to analyze the particular behaviour of some specific driveline configurations.

Even with all the advantages of a well documented and maintained general simulation program, this approach proved to cause unacceptable time delays for some reasons shown:

- The conventional computing cycle with model creation, input preparation, analysis and postprocessing is time-consuming and needs extensive user training in program usage and program handling.
- No standard elements for driveline components with special behaviour like engine, clutch or differential gear (locked or unlocked) and more. Such components had to be described in mathematical terms and coded as subroutines or procedures. This is a specialists job and well beyond the scope of designers and test engineers.
- The mixture of model definition data and mathematical code together with program execution instructions (and sometimes poor pre and postprocessing capabilities) results in long computing cycles and lags flexibility for fast design changes and parametric studies.

So it was decided to investigate the possibility to build up a specific code. Always in order to cut down the development times, at the end of an investigation phase it was decided to develop a menue driven program with integrated pre and postprocessing capabilities.

As today software and software engineering tools for nearly every hardware are available on the market, PC based software tools proofed to be exellent and could often not be found on other kind of machines. So the driveline code was specified, developed and implemented with Turbo Pascal and Turbo Tools on a PC.

It must be highlighted that the decision was based only taking into account the time needed to develop a fast user friendly application and not to reduce costs.

2.2 HARDWARE

The hardware choice has been derived from the decision taken on the software. The only consideration we may give here is, that there are noteworthy differences to work on a PC, rather than to exploit a workstation during a development phase of a new software.

The computing speed of PC's equipped with a mathematical coprocessor and a clock cycle above 20 MHz is suficient for the driveline simulation proposed here. Also the absence of CAD interfaces is not a serious lag due to the very special data needed for the simulation process.

3. ENGINEERING

Since some driveline components had to be modeled via equations anyway, it was a strategic desicion to analyze the problem in detail and write the total equation set needed for the drive train motion and the coupled vehicle interactions.

This approach requires a careful analysis of driveline component behaviour and methods to model the system with a minimum of effort but taking into account to model all effects of measurable influence. It will lenghten the design phase but it will lead to a deep understanding of the reality.

Furthermore inter department team work between the design department, engineering department and the test department is encouraged and intensified. This will avoid long periods in tuning the software to experimental data later; and last not least the end user gets the tailored application he needs.

To simulation specialist it may be boring to reproduce the set of equations of motion here. But as a matter of record the system has the following characteristics:

- The system was modeled into 10 blocks describing the driveline and vehicle components needed.
- Generalized coordinates were used to remove kinematic relations and couple all the components to a compact sytem with 14 independent degrees of freedom.
- An Adams Bashford predictor or a Runge Kutta 4 th order scheme may be used for integration in the time domain; the Givens method for real symetric eigen value extraction is used in case of normal mode analysis.
- After each integration time step forces and torques for all components are calculated using the kinematic equations to expand the condensed system of equations.
- The system was validated with 3 different sets of experimental data, consisting of engine speed and propshaft torque over time, measured during 'drop clutch tests' on different 4X2 vehicles.

4. APPLICATION

4.1 USER SHELL

All programs developed for driveline analysis, pre and postprocessing, were imbeded in an interactive menue driven user shell program named MIDAS. (Menues for Integrated Driveline AnalysiS)

MIDAS provides an graphics and printing environment and guides the user by a set of main menues and the related sub menues through all his activities.

4.2 TIME DOMAIN ANALYSIS

The time analysis program consists of a number of predefined models for the engine, clutch, gearbox, propshaft, differential gear, rear axles and wheelrims with tires. All model data from selectable databases may be changed interactive via the according menues during analysis.

Time dependent accelerations, velocities, torsions and arising torques and forces for driveline components and the vehicle suspensions due to selectable engine input are calculated and recorded as interactive selectable graphs or snapshots during a driveline session. Graphs or snapshots may be stored to disk and accessed later by the user shell MIDAS for postprocessing or documentation.

4.3 NORMAL MODE ANALYSIS

The normal mode analysis program uses the same basic modules and menue features for driveline and vehicle components as they are used by the time analysis program. Matrix methodes are used to form the eigen value problem from the selected module input data.

Natural frequencys and related mode shapes for the condesed system are calculated online by eigen value extraction and may be reviewed or recorded by snapshots selected during the analysis.

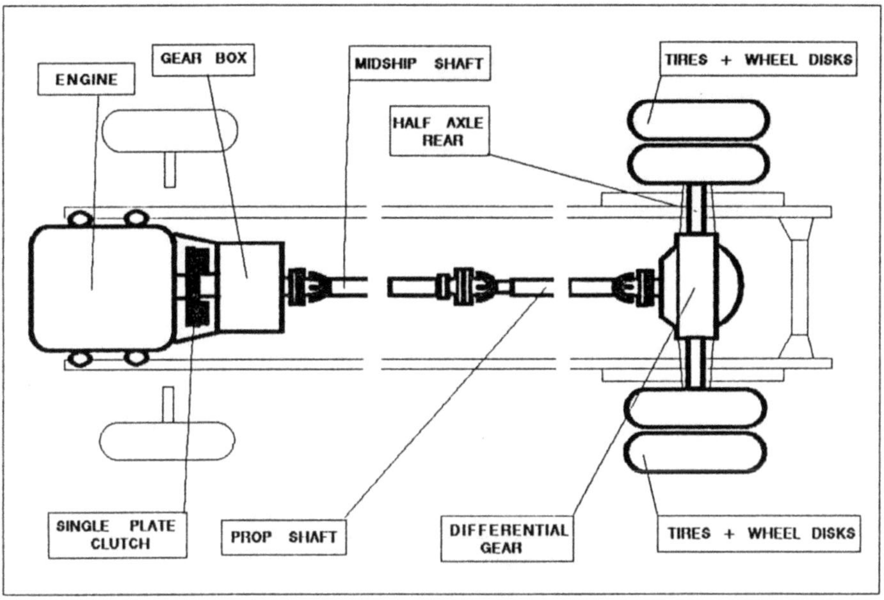

Fig. 1: DRIVELINE COMPONENTS (4 X 2 VEHICLE)

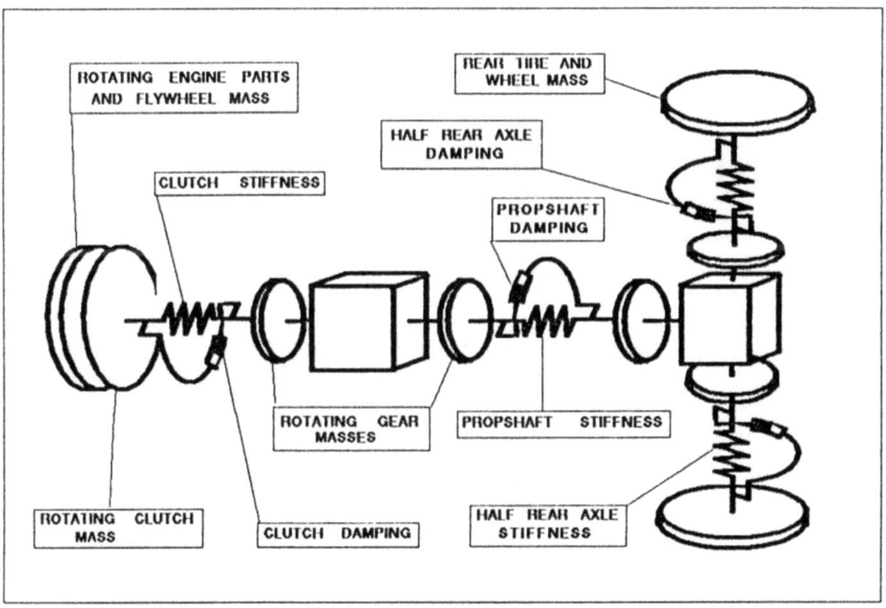

Fig. 2: LOCATION OF MASSES, SPRINGS AND DAMPERS ON DRIVE TRAIN

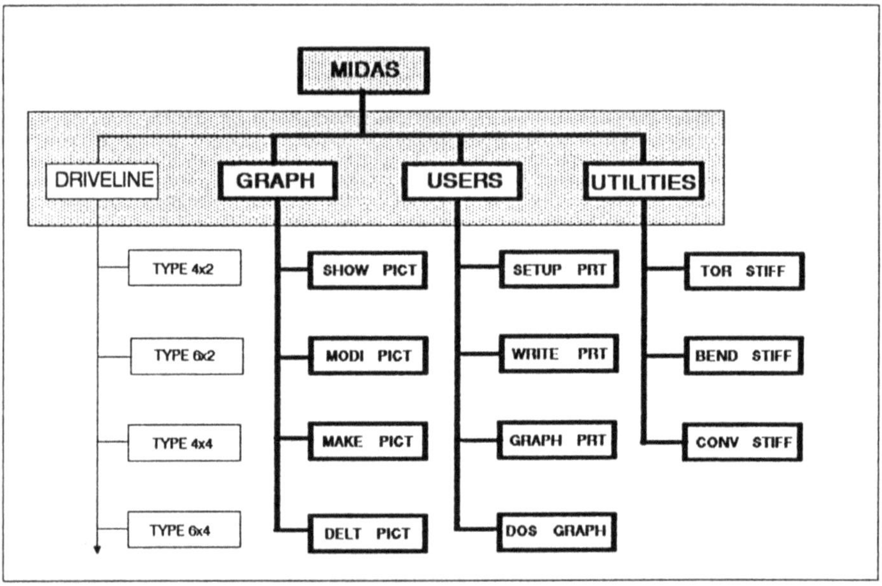

Fig. 3: DRIVELINE USER SHELL UTILITIES

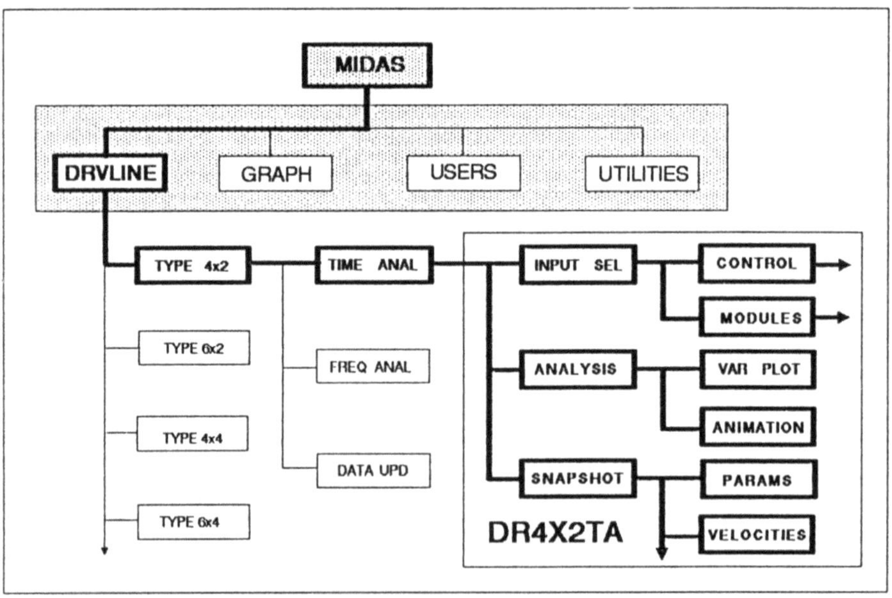

Fig. 4: USER SHELL MENUE FOR TIME ANANLYSIS

A Simulation Model for Passive Suspension Ride Performance Optimization

O.A . Olatunbosun & J.W. Dunn

Automotive Engineering Centre
University of Birmingham, U.K.

Summary

A generalised vehicle model is derived for use in the simulation of vehicle ride for optimisation of vehicle ride performance under both steady-state and transient conditions. This model allows for independent inputs at all four wheels as well as allowing for linear and non-linear elements in the suspension. The ride performance under steady-state conditions is assessed in terms of the ride discomfort parameter which is the weighted root-mean-square of the body acceleration. Under transient conditions, (i.e. traversing obstructions) the ride performance is assessed in terms of the vibration dose resulting from traversing the obstruction. Some results are presented showing the dependence of vehicle ride performance on the characteristics of various elements of the vehicle suspension and how the best combination of these may be obtained in a practical situation. The simulation model is validated using data obtained from actual vehicle tests.

Introduction

In spite of recent advances in active suspension technology, vehicles with passive suspension system are likely to dominate high volume passenger car production for the forseeable future. It is therefore important that tools are available to the suspension designer for getting the best out of the passive suspension system through optimisation of its performance. The quarter-car model has been used quite successfully in assessing the ride performance of vehicles[1,2]. However, the quarter-car model disregards pitching and rolling motions which may be important, particularly during the traversing of obstructions such as road bumps and pot-holes. The model described here is that of a full-car model.(Fig 1) which allows for independent vertical inputs at each of the four wheels and is thus able to simulate both pitching and rolling motions due to road inputs. The model is general enough to represent the full car and may include other elements not shown. The representation of the suspension system may also be as complex as desired, for example to include non-linear behaviour of elastomeric bushes. Indeed, the size of the model is limited only by the computer's memory and processing power. The model may, of course, be reduced to a half-car or quarter-car model by making appropriate assumptions.

The ride performance of a vehicle may be assessed in terms of the discomfort felt by the passenger. The root-mean-square body acceleration response to realistic random inputs from the road has been used for assessing vehicle ride discomfort[1,2,3]. This is well defined for the steady-state situation where the root-mean-square body acceleration is a measure of the vibration power absorbed by the subject. Attempts have been made to correlate subjective assessment of vehicle ride discomfort with objective measurement of body acceleration by applying various weighting functions[4]. Using the results of a large number of experiments, the ISO has produced weighting functions for equalising the discomfort produced by vibrations at different frequencies[5]. These have been used to define a Discomfort parameter for assessing vehicle ride performance in a steady-state situation[2] which is the root-mean square weighted body acceleration. In a transient situation, an average quantity such as the r.m.s. value is less useful than a cumulative (i.e. dose) measure. The vibration dose value[4] is a cumulative measure which has been used to assess the severity of ride with respect to annoyance and comfort. Hence, the vibration dose value, defined as the root-mean-quad of the body acceleration, with respect to time, in traversing the obstruction, is adopted as the measure of ride discomfort in this case and reflects the vibration dose transmitted to the body.

Methods of solution of the model for both steady-state road inputs and transient inputs are outlined. Results of simulations are presented to illustrate the dependence of vehicle ride performance on the characteristics of various elements of the vehicle suspension and how the best combination of these may be obtained in a practical situation. The simulation results are validated using data from actual vehicle tests.

Derivation of the Vehicle Model

A full car model is shown in Figure 1. This is a considerably simplified model of a car in which the body structure is regarded as rigid and simple two degrees-of-freedom suspension models are used. The assumption of a rigid body structure is valid as consideration of ride is limited to low frequencies below 20 Hz. The suspension model is very simple but even with the very simple model, the derivation of the equations of motion can get rather involved. However, a generalised model, derived below, has been implemented in a computer program to generate the system equations of motion automatically from user input data. The program has been implemented on a personal computer and has coped successfully with up to 20 degrees of freedom. However, the model is really not limited in terms of degrees of freedom, only in terms of the computer hardware.

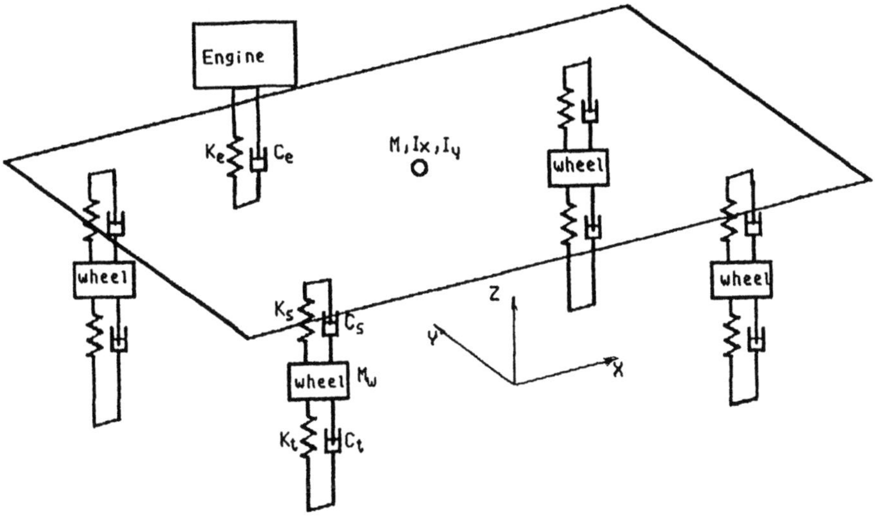

Fig 1 **Generalised Vehicle Model**

For such a system as shown in Fig 1, the equations of motion may be derived from Lagrange's equation[6] i.e.

$$\frac{d}{dt}\left[\frac{\partial T}{\partial \dot{q}}\right]-\frac{\partial T}{\partial T}+\frac{\partial D}{\partial \dot{q}}+\frac{\partial V}{\partial q}=F_q \tag{1}$$

where T,D and V represent the system kinetic, dissipative and potential energies respectively. T is the sum of the kinetic energies of the individual masses and inertias i.e.

$$T=\sum_1^n m_i \dot{x}_i^2=\frac{1}{2}\{\dot{X}\}^T[diag(m)]\{\dot{X}\} \tag{2}$$

where

$$[diag(m)]=\begin{bmatrix} m_1 & 0 & \cdots & 0 \\ 0 & \ddots & & \vdots \\ \vdots & & \ddots & 0 \\ 0 & \cdots & 0 & m_n \end{bmatrix} \tag{3}$$

and

$$\{X\}=\begin{Bmatrix} x_1 \\ \vdots \\ \vdots \\ x_n \end{Bmatrix} \tag{4}$$

D and V are the sums of the energies in the dampers and springs respectively, i.e.

$$D = \sum_{1}^{n} \frac{1}{2} c_i \dot{y}_i^2 = \frac{1}{2} \{\dot{Y}\}^T [diag(c)] \{\dot{Y}\} \tag{5}$$

and

$$V = \sum_{1}^{n} \frac{1}{2} k_i y_i^2 = \frac{1}{2} \{Y\}^T [diag(k)] \{Y\} \tag{6}$$

where

$$[diag(c)] = \begin{bmatrix} c_1 & 0 & \cdots & 0 \\ 0 & \ddots & & \vdots \\ \vdots & & \ddots & 0 \\ 0 & \cdots & 0 & c_n \end{bmatrix} \tag{7}$$

$$[diag(k)] = \begin{bmatrix} k_1 & 0 & \cdots & 0 \\ 0 & \ddots & & \vdots \\ \vdots & & \ddots & 0 \\ 0 & \cdots & 0 & k_n \end{bmatrix} \tag{8}$$

and

$$\{Y\} = \begin{Bmatrix} y_1 \\ \vdots \\ y_n \end{Bmatrix} \tag{9}$$

$y_1, y_2, \text{--} y_n$ are the displacements across (i.e. extensions in) the springs and dampers.

The vector {Y} has the following relationship:-

$$\{Y\} = [H]\{X\} - [G]\{U\} \tag{10}$$

where

$[H]$ and $[G]$ are associative matrices

and

$\{U\}$ is a vector representing externally applied displacements e.g. at the tyre/road interface

Using the above relationships, it can be shown that

$$V = \frac{1}{2}\{X\}^T [H]^T [diag(k)][H]\{X\} - \{X\}^T [H]^T [diag(k)][G]\{U\}$$
$$+ \{U\}^T [G]^T [diag(k)][G]\{U\} \tag{11}$$

i.e.

$$V = \frac{1}{2}\{X\}^T [K]\{X\} - \{X\}^T [K_1]\{U\} + \frac{1}{2}\{U\}^T [K_2]\{U\} \tag{12}$$

where

$$[K]=[H]^T[diag(k)][H] \quad (13a)$$

$$[K_1]=[H]^T[diag(k)][G] \quad (13b)$$

and

$$[K_2]=[G]^T[diag(k)][G] \quad (13c)$$

Likewise

$$D=\frac{1}{2}\{\dot{X}\}^T[H]^T[diag(c)][H]\{\dot{X}\}-\{\dot{X}\}^T[H]^T[diag(c)][G]\{\dot{U}\}$$
$$+\{\dot{U}\}^T[G]^T[diag(c)][G]\{\dot{U}\} \quad (14)$$

i.e.

$$D=\frac{1}{2}\{\dot{X}\}^T[C]\{\dot{X}\}-\{\dot{X}\}^T[C_1]\{\dot{U}\}+\frac{1}{2}\{\dot{U}\}^T[C_2]\{\dot{U}\} \quad (15)$$

where

$$[C]=[H]^T[diag(c)][H] \quad (16a)$$

$$[C_1]=[H]^T[diag(c)][G] \quad (16b)$$

and

$$[C_2]=[G]^T[diag(c)][G] \quad (16c)$$

Equation (2) may be written as:-

$$T=\frac{1}{2}\{\dot{X}\}^T[M]\{\dot{X}\} \quad (17a)$$

where

$$[M] = [diag(m)] \quad (17b)$$

Substituting equations (12), (15) and (17) into the Lagrange's equation (1) gives the equation of motion:-

$$[M]\{\ddot{X}\}+[C]\{\dot{X}\}+[K]\{X\}=\{F\} \quad (18a)$$

where

$$\{F\}=[K_1]\{U\}+[K_2]\{\dot{U}\} \quad (18b)$$

$[M]$, $[K]$ and $[C]$ represent the system mass, damping and stiffness matrices respectively.

The process of formulating the system matrices $[M]$, $[K]$, $[C]$, $[K_1]$ and $[C_1]$ may be automated to allow the generalisation of the system model. This has been implemented as a computer program which accepts system parameters either in the form of input files or interactively, for a model having up to 20 degrees of freedom. Hence, the system model may be generated quickly and easily.

The solution of the system model may then be obtained using the appropriate input functions for the vectors $\{\dot{U}\}$ and $\{U\}$.

Solution for Steady state road input

Analysis of the steady-state behaviour is based on road inputs described with a displacement mean square spectral density function of the form $S(\nu)=\frac{\kappa}{\nu^{2.5}}$ m^3/cycle where κ is a roughness constant and ν is the wave number[6]. The ride performance is assessed in terms of the a *discomfort parameter* which is the root-mean-square of the body acceleration i.e.

$$discomfort = \sqrt{\int_0^{\infty} a^2(w)d\omega} \tag{19a}$$

The rapid convergence of the integral means that only the frequency range of 1 to 20 Hz need be considered. The body acceleration is weighted appropriately using the ISO[4] fatigue reduced comfort boundary. i.e.

$$discomfort = \sqrt{\int_0^{\infty} a^2(\omega)W(\omega)d\omega} \tag{19b}$$

If the input is harmonic then the system equation may be written as:

$$\left[-\omega^2[M]+j\omega[C]+[K]\right]\{X\} = \left[[K_1]+j\omega[C_1]\right]\{U\} \tag{20a}$$

or

$$[Y]\{X\} = [Y_1]\{U\} \tag{20b}$$

where

$$[Y] = -\omega^2[M]+j\omega[C]+[K] \tag{20c}$$

and

$$[Y_1] = [K_1]+j\omega[C_1] \tag{20d}$$

Hence:-

$$\{X\} = [Y]^{-1}[Y_1]\{U\} \tag{21}$$

The value of the body acceleration at each frequency is obtained from the vector $\{X\}$ for substitution into the equation (19) to obtain the discomfort parameter..

Analysis of Transient Response

The transient response of the system to traversing an obstruction such as a road bump or pot-hole may be obtained by piece-wise numerical integration, allowing realistic system behaviour, including suspension 'bottoming' and loss of road/wheel contact, to be simulated. The body acceleration in this case is transient and cannot be described in the same terms as for the steady-state case. The ride performance is therefore assessed in this case in terms of the *vibration dose value* which is a cumulative measure of the vibration transmitted to the body in traversing the obstruction. This parameter is defined as the root-mean-quad of the body acceleration, with respect to time , in traversing the obstruction.

Equation (18) may be written as:

$$\{\ddot{X}\}+[A_1]\{\dot{X}\}+[A_2]\{X\}=[A_3]\{\dot{U}\}+[A_4]\{U\} \tag{22a}$$

where

$$\begin{aligned}[A_1]&=[M]^{-1}[C]\\ [A_2]&=[M]^{-1}[K]\\ [A_3]&=[M]^{-1}[C_1]\\ [A_4]&=[M]^{-1}[K_1]\end{aligned} \tag{22b}$$

Dropping the matrix notation, the equation (22a) may be written as:-

$$\dot{Z}+A_1Z+A_2X=A_3\dot{U}(t)+A_4U(t) \tag{23}$$

where

$$Z=\dot{X} \tag{24}$$

The resulting equation is thus a first-order differential equation for which a solution may be obtained by a suitable numerical method. Here, the Runge-Kutta method[6,7] is used. The vectors, $\dot{U}(t)$ and $U(t)$ are obtained by discretising the road input functions resulting from the interaction of the tyre with the obstruction using small time steps. The response of the system is thus computed at each time step. The system response at time t+Δt is given by:-

$$Z(t+\Delta t)=Z(t)+\frac{1}{6}\left[\beta\ +2\beta_2+2\beta_3+\beta_4\right] \tag{25a}$$

$$X(t+\Delta t)=X(t)+\frac{1}{6}\left[\alpha_1+2\alpha_2+2\alpha_3+\alpha_4\right] \tag{25b}$$

where

$$\begin{aligned}\beta_1&=\Delta t\left[A_3\dot{U}(t)+A_4U(t)-A_1Z(t)-A_2X(t)\right]\\ \alpha_1&=\Delta t[Z(t)]\end{aligned} \tag{26a}$$

$$\begin{aligned}\beta_2&=\Delta t\left[A_3\dot{U}(t+\frac{\Delta t}{2})+A_4U(t+\frac{\Delta t}{2})-A_1(Z(t)+\frac{\beta_1}{2})-A_2(X(t)+\frac{\alpha_1}{2})\right]\\ \alpha_2&=\Delta t\left[Z(t)+\frac{\beta_1}{2}\right]\end{aligned} \tag{26b}$$

$$\begin{aligned}\beta_3&=\Delta t\left[A_3\dot{U}(t+\frac{\Delta t}{2})+A_4U(t+\frac{\Delta t}{2})-A_1(Z(t)+\frac{\beta_2}{2})-A_2(X(t)+\frac{\alpha_2}{2})\right]\\ \alpha_3&=\Delta t\left[Z(t)+\frac{\beta_2}{2}\right]\end{aligned} \tag{26c}$$

$$\begin{aligned}\beta_4&=\Delta t\left[A_3\dot{U}(t+\Delta t)+A_4U(t+\Delta t)-A_1(Z(t)+\beta_3)-A_2(X(t)+\alpha_3)\right]\\ \alpha_4&=\Delta t\left[Z(t)+\beta_3\right]\end{aligned} \tag{26d}$$

The acceleration at time t is given by:-

$$\dot{Z}(t) = A_3\dot{U}(t) + A_4U(t) - A_1Z(t) - A_2X(t) \qquad (27)$$

The vibration dose value is defined as the root-mean-quad of the body acceleration [4] and is given by:

$$vib\_dose = \left[\int_0^T a^4 dt\right]^{0.25} \qquad (28)$$

Simulation Results and Optimisation

The simulation results reported here are based on a mini-bus vehicle. Several models of varying complexity the were used in the simulations. The results of the more complex models were compared to those of the simpler models to see if anything was gained in increasing the complexity of the model. In the case of the steady-state analysis, it was shown that a quarter-car model was quite adequate for investigating the effects of the main suspension parameters. However, in the analysis of the traversing of obstructions, it was shown that there was significant difference between results obtained from the quarter-car model and more complex models. The results shown here have been obtained from a full-car model of the type shown in Figure 1. No attempt has been made as yet to include the non-linear effects due to elastomeric mounts.

Figure 2 shows the simulated body acceleration response for the base model i.e. using the measured vehicle parameters, in traversing a half-sine bump. Also shown is the measured response of the actual vehicle to a simulated half-sine bump. Comparison of the theoretical and measured accelerations indicate the good accuracy of the model.

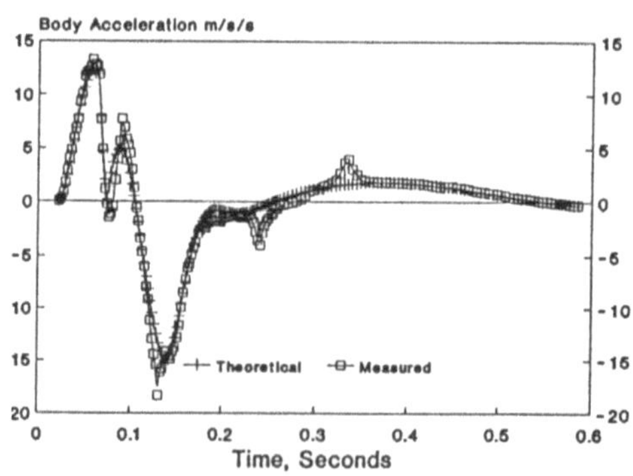

Fig 2: Body Acceleration Response; Half-sine bump

Whilst minimising passenger discomfort is a prime objective of the optimisation exercise, certain constraints are imposed by the need for the suspension system to maintain even tyre loading for vehicle attitude control in accelerating, braking and cornering. The dynamic tyre load must therefore be limited in order for the vehicle to be able to maintain steering control. It must also achieve this within a limited wheel travel. Hence, suspension working space must be limited to a reasonable value. Optimisation of the suspension therefore requires that both dynamic tyre load and suspension working space be maintained within reasonable limits while striving to minimise the discomfort parameter. In the simulations, both dynamic tyre load and suspension wheel travel are determined in addition to the discomfort parameter. The results

Figure 3.1 shows the variation of the vehicle ride parameters with the suspension stiffness. for the base model. The simulation is for a speed of 30 km/hr on an average rough road for obtaining the discomfort parameter, dynamic tyre load and suspension working space. The vibration dose value is for traversing a half-sine bump of height 0.1m and width 0.5m. Only the dynamic tyre load shows considerable variation with suspension stiffness.

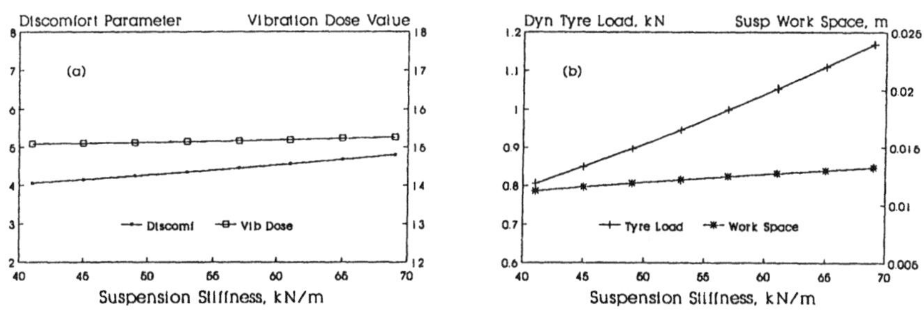

Fig 3.1: Ride Parameters vs Suspension Stiffness

Figure 3.2 shows the variation of the vehicle ride parameters with suspension damping also for the base model. All parameters vary considerably with the suspension damping. While the vibration dose increases with increasing damping, the discomfort seems to have a minimum value at a damping value which also gives a low value of vibration dose. However, the suspension working space is quite high with this value of damping and the minimum dynamic tyre load is also displaced from this value. An optimisation exercise will attempt to determine the combination of vehicle suspension parameters which will produce minimum discomfort while also minimising the dynamic tyre load and keeping the working space within reasonable limits.

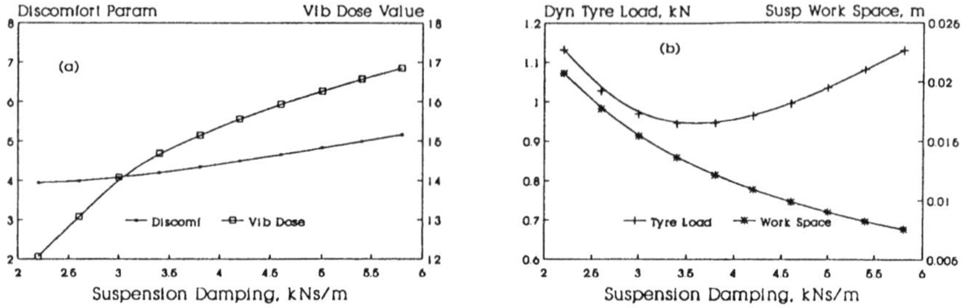

Fig 3.2: Ride Parameters vs Suspension Damping

Fig 3.3 shows the variation of the ride parameters with tyre stiffness for the base model. The vibration dose is particularly sensitive to tyre stiffness increasing rapidly as the tyre stiffness increases. Ride discomfort also increases as the tyre stiffness increases, but not by very much. Both the dynamic tyre load and suspension working space decrease as tyre stiffness increases.

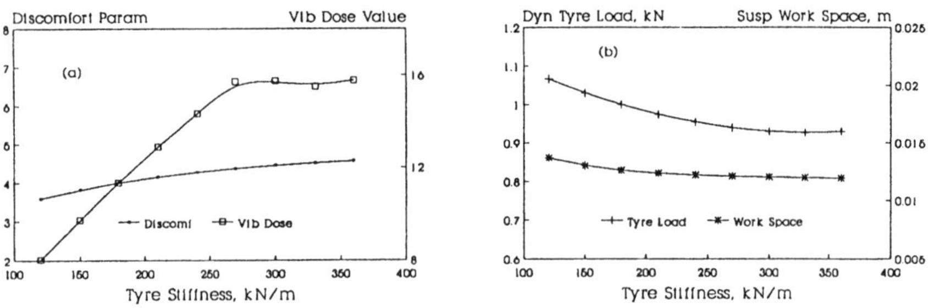

Fig 3.3: Ride Parameters vs Tyre Stiffness

Figure 4 shows the effect of varying both the suspension stiffness and damping to try and obtain the best combination of ride parameters. The results indicate that a value of suspension damping between 3 and 3.5 kNs/m would give the best ride assuming that all other parameters are fixed. For the suspension stiffness, avalue of between 41kN/m and 45kN/m is indicated. This gives a discomfort parameter of less than 4 and a vibration dose of approximately 14 while minimising the dynamic tyre load at approximately 0.8 kN and keeping the working space within reasonable limits. This may be improved by optimising the other parameters as well.

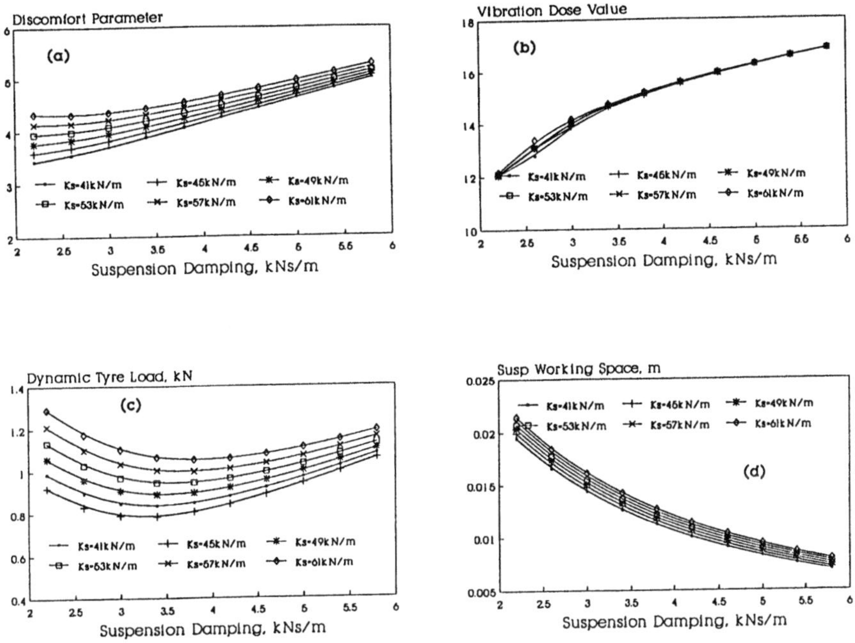

Fig 4: Ride Parameters: Variation of Suspension Stiffness and Damping

Conclusions

1. A generalised vehicle model has been derived for use in the simulation of vehicle ride.
2. The derived model has been used for simulating both steady-state and transient ride conditions with the objective of optimising the vehicle parameters for ride performance.
3 The ride discomfort parameter has been used to assess steady-state ride performance while the vibration dose value has been used for transient ride performance.
4. Results presented indicate the good accuracy of the model and show the dependence of the ride performance on the various vehicle suspension parameters. An example of how the ride performance may be optimised is shown.

References

1. Thompson, A.G.; Optimum damping in a randomly excited non-linear suspension. *Proceedings Instn Mech Engnrs*. vol 184, pt 2A, no 8 (1969-70) pp169-184.

2. Sharp, R.S. & Hassan, S.A.; An evaluation of passive automotive suspension systems with variable stiffness and damping parameters. *Vehicle System Dynamics,* 15 (1986) pp335-350.

3. Abdel Hady, M.B.A. & Crolla, D.A.; Theoretical analysis of active suspension performance using a four-wheel vehicle model. *Proc Instn Mech Engr*, 203 (1989), part D pp125-135.

4. Griffin, M.J.; Evaluation of vibration with respect to human response. *SAE paper* no 860047.

5. I.S.O. 2631/1 1985.; Evaluation of human exposure to whole-body vibration - Part 1: General requirements.

6. Thomson, W.T.; Vibration Thoery and applications, 3rd Edition. *George Allen &Unwin*, London 1988.

7. Hamming, Numerical methods for engineers and scientists. *McGraw-Hill,* London 1962

Computational Fluid Dynamics

Development Issues for the ARINS Algorithm for Vehicle Passenger Compartment Flow Simulation

A. J. Baker and P. D. Manhardt
Computational Mechanics Corp. (COMCO)
Knoxville, TN 37919 USA

Abstract

Safety and comfort considerations demand that vehicle passenger compartment Heating/Ventilating/Air Conditioning (HVAC) systems be optimized to promote a quality environment with efficiency. Vehicle HVAC systems are characterized by multiple supply outlets, variable compartment exhausts, radiation loading and significant flow obstructions including passengers. Theoretical and computational issues are summarized for the ARINS finite element algorithm being developed for the problem class.

INTRODUCTION

Significant marketing leverage accrues, in the highly competitive automotive (autos, trucks, buses) industry, to passenger comfort optimization. This is evidenced by the attention now being paid to cabin ergonomics, i.e., instrument displays (including head-up), acoustic insulation, multiple quadraphonic CD sound systems, posture-adjustable seats, etc. As cabin interiors become "sexier," the designer must also focus on passenger *total* thermal comfort, including those occupying the back seat! Opportunity certainly exists for innovative HVAC design, e.g., ducting conditioned air through glass headers and/or door sills, for promoting an acceptable transient from hot/cold soak. The design issue is thus safety and comfort, with operating economy, which translates into vehicle "interior environment quality" with efficiency.

The authors over several years have examined theory and code design issues associated with construction of a mathematical, hence computational fluid dynamics (CFD), model capable of accurately and efficiently addressing all the pertinent issues. The governing conservation law (partial differential equation, PDE) system is Reynolds-averaged, incompressible-thermal Navier-Stokes. A turbulence closure model is required, radiation loading can be a significant factor, and transport of an inert species may be important. The HVAC supply side has multiple vaned outlets, and flow exhausts from passenger compartments through diverse locations including "leaks." Total flow redirection can be obtained via supply vane orientation, and/or modest pressure differentials, e.g., slightly opening a window. A steady-state determination is important, however, transient evolution from a cold/hot soak to operating steady-state addresses a demanding problem related to comfort and safety. Finally, the flow geometries associated with automotive passenger compartments are non-aerodynamic and multiply-connected.

Significant progress since the 2nd European conference [1] has achieved the goal of a mathematically robust and computationally stable CFD simulation theory and code. This paper summarizes key elements of the ARINS finite element (FE) CFD algorithm developed for automotive HVAC analyses. Research codes have produced quality solutions to critical benchmark problems, discussed herein, which point the way to design of the ARINS algorithm "template" for *AKCESS.*™*, COMCO's developing UNIX production shell, for vehicle HVAC system design optimization.

THE PROBLEM STATEMENT

Any of several statistical manipulations of the incompressible Navier-Stokes equations yields "*Reynolds-averaged*" Navier-Stokes (RINS) systems governing turbulent, incompressible thermal flows. Following a suitable non-dimensionalization, and assuming a turbulent kinematic "eddy" viscosity closure model, the simplest form for RINS is

$$\mathcal{L}(\rho_0) = \nabla\cdot\mathbf{u} = 0 \tag{1}$$

$$\mathcal{L}(\mathbf{u}) = \frac{\partial\mathbf{u}}{\partial t} + (\mathbf{u}\cdot\nabla)\mathbf{u} - \nabla\cdot\left(\frac{1}{\mathrm{Re}} + \nu^t\right)\nabla\mathbf{u} + \nabla P + \frac{\mathrm{Gr}}{\mathrm{Re}^2}\Theta\hat{\mathbf{g}} = 0 \tag{2}$$

$$\mathcal{L}(\Theta) = \frac{\partial\Theta}{\partial t} + (\mathbf{u}\cdot\nabla)\Theta - \nabla\cdot\left(\frac{1}{\mathrm{RePr}} + \frac{\nu^t}{\mathrm{Pr}^t}\right)\nabla\Theta - s_\Theta = 0 \tag{3}$$

$$\mathcal{L}(Y_A) = \frac{\partial Y_A}{\partial t} + (\mathbf{u}\cdot\nabla)Y_A - \nabla\cdot\left(\frac{1}{\mathrm{ReSc}} + \frac{\nu^t}{\mathrm{Sc}^t}\right)\nabla Y_A - s_A = 0 \tag{4}$$

The dimensionless groups in (2)-(4) have the definitions

$$\mathrm{Re} \equiv \frac{UL}{\nu}, \quad \mathrm{Gr} \equiv \frac{g\,\beta(T_{max} - T_{min})\,L^3}{\nu^2}, \quad \mathrm{Pr} \equiv \frac{\rho_0 c_p \nu}{k}, \quad \mathrm{Sc} \equiv \frac{\nu}{D_{AB}} \tag{5}$$

where Re is Reynolds number, Gr is the Grashoff number, Pr is Prandtl number and Sc is the Schmidt number. A superscript "t" on any variable denotes its modeled "turbulent" counterpart.

The dependent variable set in (1)-(3) includes mean flow velocity vector $\mathbf{u}(\mathbf{x},t)$ with scalar resolution u_i, $1 \leq i \leq n$ in n-dimensional space, the kinematic pressure ($P=p/\rho_0 + 2k/3$) and potential temperature $\Theta \equiv (T - T_{min})/(T_{max} - T_{min})$. In (4), Y_A is the mass fraction of an inert species A, for example, a smoke tracer, and D_{AB} is the binary diffusion coefficient. Finally, the Boussinesq body force approximation for buoyancy is made in (2) and $\hat{\mathbf{g}}$ is the gravity unit vector.

The mathematical characterization of (1)-(4) depends upon these non-dimensional groups. The highest order spatial derivatives exist in (2)-(4), hence each isolated equation is elliptic for finite Reynolds number Re. Therefore, knowledge of the mean flow variables, their normal derivatives, and/or linear combinations of these data is required everywhere on the boundary $\partial\Omega$ of the n-dimensional spatial domain of definition $\Omega \subset \Re^n$. In distinction, Re goes to infinity as $\nu \rightarrow 0$ in the inviscid flow assumption, and (1)-(3) becomes of hyperbolic conservation law form. Finally, for constant density ρ_0, the continuity equation (1) dominates the mathematical well-posedness issue for any proposed CFD theory.

Specifically, (1) defines the *differential constraint* that velocity fields $\mathbf{u}(\mathbf{x}, t)$ admissible as solutions to (2)-(3) must be divergence-free. This constraint ultimately dominates the pressure (gradient) field acting as a source in (2). For incompressible (or constant density) fluids, neither an algebraic equation of state nor a differential equation is available for determination of pressure. Thus, pressure can be eliminated mathematically only by imposing a dependent variable transformation to streamfunction and vorticity vectors.

The alternative is to develop an inexact, i.e., "CFD," theory such that pressure effects become adequately "simulated" during a convergent iterative process. The three mathematically distinct formulations are constituted of, 1) an algebraic equation, 2) an elliptic differential equation, and 3) an initial-value hyperbolic differential equation. Each theory possesses distinct features and consequential detractions. Following thorough assessment, an inexact iteration involving a Poisson equation appears preferable for 3-dimensional simulations. In 2 dimensions, this CFD theory performance can be critically benchmarked against the exact streamfunction-vorticity formulation, which is robust in 2-D only (unfortunately!).

Beside pressure, the other issue central to closure for (1)-(4) is the turbulence model, from which the distribution of kinematic "eddy viscosity" ν^t is determined. The "industry standard" for RINS systems has been the turbulent kinetic energy-isotropic dissipation (k-ε) model, which introduces an additional pair of PDEs into (1)-(4). For finite Re, this isolated PDE set is elliptic, hence requires appropriate data specification all around the boundary $\partial\Omega$ of Ω. The fact that ν^t, hence turbulence Reynolds number $\mathrm{Re}^t \equiv \nu^t/\nu$, goes sharply to zero in regions directly adjacent to solid boundaries, places very severe mesh resolution demands for boundary condition implementation. The usual CFD "cure" is use of boundary layer *similarity* concepts, to replace the exact statement with an approximation displaced some distance away from a wall. Use of such "wall-function" boundary conditions can consequentially compromise accuracy, especially around flow separation regions, which is the price extracted for not resolving wall-layer distributions.

Turbulence closure modeling should ultimately be the *only* "weak link" in a CFD simulation for RINS. As currently practiced, many (most) CFD theory/code implementations augment laminar (ν) and turbulent (ν^t) diffusion effects with "numerical diffusion," which can be interpreted as an artificial viscosity ν^a. In the spirit of the turbulence Reynolds number definition, one could also define the artificial diffusion Reynolds number $\mathrm{Re}^a \equiv \nu^a/\nu$. Then, since one also usually assumes that $\mathrm{Pr}^t \approx \mathrm{Pr}$ and $\mathrm{Sc}^t \approx \mathrm{Sc}$, in closing (2)-(4), it is possible to extract common multipliers in (2)-(4) yielding, for example, for (2)

$$\left(\frac{1}{\mathrm{Re}} + \nu^t\right) \Rightarrow \frac{1}{\mathrm{Re}}\left(1 + \mathrm{Re}^t + \mathrm{Re}^a\right) \tag{6}$$

Equation (6), and the corresponding expressions for (3)-(4), leads to the concept of an "effective" Reynolds number Re^f defined as

$$\mathrm{Re}^f \equiv \frac{\mathrm{Re}}{1 + \mathrm{Re}^t + \mathrm{Re}^a} \tag{7}$$

Clearly, the least diffusive characterization corresponds to a laminar flow ($\mathrm{Re}^t=0$) without added artificial viscosity. Thereafter, $\mathrm{Re}^t>0$ and/or $\mathrm{Re}^a>0$ serves to increase diffusive effects, and their individual actions can be indistinguishable using some current production codes. For the desired basic mathematical characterization of candidate CFD theories, and codes, it is sufficient to control Re^f to quantitatively study the interplay between approximation diffusive and dispersive error mechanisms.

FINITE ELEMENT CFD ALGORITHM

Of the various inexact theories for imposing (1) within an interactive CFD strategy, the observation that the attendant error in $\mathbf{u}^p$, a (any) approximation to the solution $\mathbf{u}$ to (2), can be irrotational leads to a Poisson equation for the associated potential function $\phi(\mathbf{x},t)$. The known boundary conditions for ϕ are (only) homogeneous Neumann, hence one must create a suitable Dirichlet constraint to render the terminal matrix statement non-singular. Assuming this resolvable, (1) is replaced by

$$\mathcal{L}(\phi) \quad = \quad -\nabla^2\phi + s\,(\mathbf{u}^p) = 0 \tag{8}$$

and (2)-(4), upon defining $\nu^f \equiv (Re^f)^{-1}$, are the hyperbolic conservation law form

$$\mathcal{L}(q) \quad = \quad \frac{\partial q}{\partial t} + \mathbf{u}\cdot\nabla q - \nabla\cdot\left(\nu_q^f\,\nabla q\right) - s_q = 0 \tag{9}$$

The RiNS state variable for (9) is $q(\mathbf{x},t) = \{\mathbf{u}, \Theta, Y_A, k, \varepsilon, \ldots\}^T$, and s_q is the pertinent source term.

An FE algorithm for (9) is a specific semi-discretization of a weak statement written for *any* approximation $q^N(\mathbf{x},t)$ to $q(\mathbf{x},t)$. The *optimal* Galerkin weak statement renders the resultant approximation error $\mathcal{L}(q^N)$ othogonal to the trial space supporting q^N, i.e.,

$$WS^N \equiv \int_\Omega \Psi_j(\mathbf{x})\,\mathcal{L}(q^N)\,d\tau \equiv 0\ ,\ \forall_j \tag{10}$$

for $q(\mathbf{x},t) \approx q^N(\mathbf{x},t) \equiv \Sigma\,\Psi_j(\mathbf{x})\,Q_j(t)$, $1 \le j \le N$. The FE formulation [2] constructs q^N using (usually) compact polynomial interpolations of each $\Psi_j(\mathbf{x})$ on the generic union of finite elements Ω_e sharing a node of the discretization, $\Omega^h \equiv \cup\Omega_e$, of the spatial domain of definition for (9). Significantly, this discretization need not employ staggered nodal degrees of freedom, as is required for most non-Galerkin weak statements, e.g., a "finite volume" algorithm.

The resulting expression for the semi-discrete Galerkin WS^N, i.e., GWS^h, for (9) produces the matrix ordinary differential equation (ODE) system

$$GWS^h \equiv [M]\frac{d\{Q\}}{dt} + \{R(Q)\} = \{0\} \tag{11}$$

In (11), $[\bullet]$ and $\{\bullet\}$ denote a square and column matrix, respectively, and $\{Q\} \equiv \{Q(t)\}$ is the array of the state variable semi-discretization at the nodes of Ω^h. Any ODE algorithm utilizes (10) to evaluate terms in a temporal Taylor series. For example, the terminal matrix algebra statement

$$\{FQ\} = [M]\left\{Q_{n+1} - Q_n\right\} + \Delta t\left(\theta\{R\}_{n+1} + (1-\theta)\{R\}_n\right) = \{0\} \tag{12}$$

results for the family of θ-implicit, one-step (Euler/trapezoidal) methods, where $t_{n+1} = t_n + \Delta t$. The complete FE solution algorithm is constituted of (12) and the GWS^h for (8), which is

$$\{F\Phi\} = [D]\left\{\Phi\right\} + \{s\} = \{0\} \tag{13}$$

where $[D]$ is the "diffusion" matrix equivalent of $(\nabla^2)^h$, the discretized laplacian operator on Ω^h.

Equations (12)-(13) constitute a non-linear algebraic system for $\theta>0$. The class of candidate numerical linear algebra solution procedures amounts to approximations of the Newton statement

$$[J]\,\{\delta Q\}_{n+1}^{p+1} = -\,\{F\}_{n+1}^{p} \tag{14}$$

where the jacobian definition is $[J] \equiv \partial\{F\}/\partial\{Q\} = [M] + \theta\Delta t\partial\{R\}/\partial\{Q\}$, and p is the iteration index at time t_{n+1}. The CFD theory suggests that solving (13), after an iteration on (14) produces updated data for $\{s\}$, is appropriate.

The available mathematical theory [3] characterizing the FE algorithm (12)-(14) predicts that a kth degree polynomial basis nominally yields a spatially $2k$ order-accurate algorithm for bounded Re for the linearized analysis. For the trapezoidal rule, $\theta=1/2$ in (11), any FE basis GWS^h algorithm is intrinsicly *free* from artificial viscosity ν^a, hence $Re^a=0$ in (6). For the explicit ODE choice $\theta=0$, $Re^a>0$, the usable Δt is limited by the classical CFL condition, and $[M]$ in (14) must be diagonalized for efficiency. Any choice $\theta>0.5$ also yields $Re^a>0$. Since Ω^h does not employ staggered degrees-of-freedom, all boundary conditions are indeed imposed on $\partial\Omega^h$.

DISCUSSION AND RESULTS

Several benchmark test cases have been selected/designed to verify the FE algorithm, (12)-(14), for key geometrical/flow issues pertinent to automotive HVAC simulation requirements. Results highlights by category follow.

Massive separation: A $2D/3D$ cartesian geometry for assessing isothermal massive flow separation, and comparison to experiment [4], is the step-wall diffuser. Figure 1a illustrates the geometry where, for a step-height/inlet duct width ratio near unity, up to four distinct separation regions become induced, as a function of Re, before the flow transitions to turbulent. The flow remains $2D$ and laminar for Re $\leq$ 400, becomes $3D$ on 400 $<$ Re $\leq$ 1200 while remaining laminar, and is fully turbulent for Re$>$8000. For the $2D$ benchmark, excellent agreement with experiment is indicated for Re $\leq$ 400, Fig. 1b. Continuing with

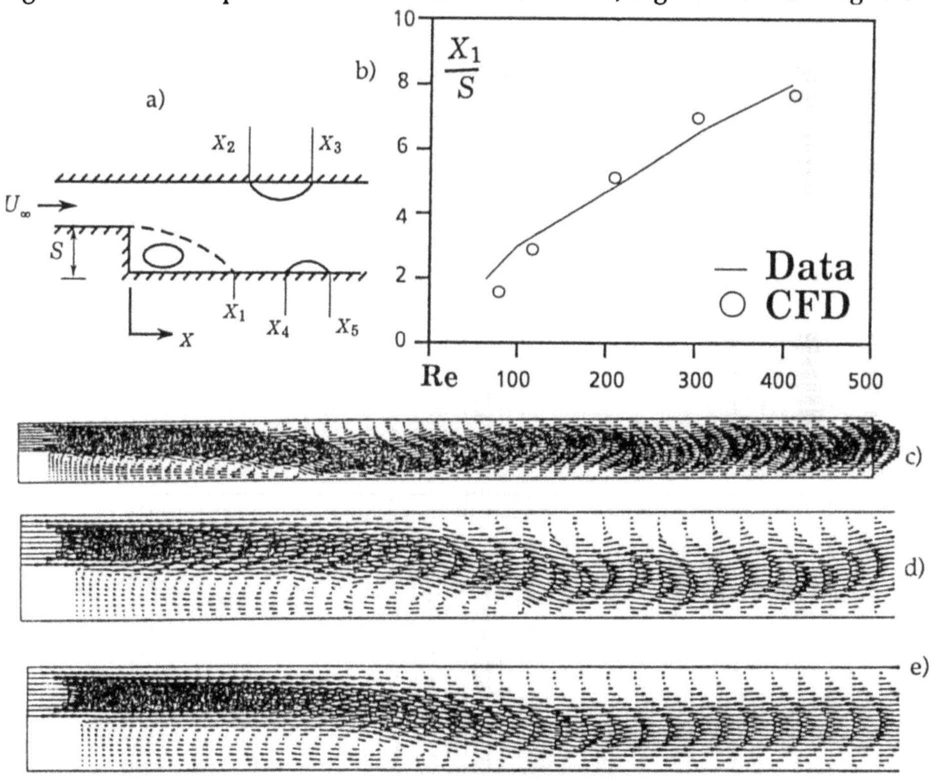

Figure 1. Step-wall diffuser benchmark, a) geometry, b) primary recirculation region, Re≤400, c) complete steady velocity field, Re=1000, d)-e) close-up comparison of constraint and CFD streamfunction solutions, Re=10^3

2D, Fig. 1c graphs the excellent quantitative comparison obtained between the inexact constraint theory, and the exact streamfunction solution, for three distinct separation regions at Re = 1000. For continuation to Re = 1200, both CFD solutions become unsteady, in qualitative agreement with the experiment showing first signs of transition to turbulent flow.

Flow bifurcation: Multiple passenger compartment exhaust locations requires that a CFD simulation exist without *a priori* knowledge of exit flow bifurcation, i.e., one cannot define the outlet velocity field. Examination of the CFD literature confirms that exhaust velocity (distribution) is typically assumed known, which also bypasses appropriate Dirichlet data for the constraint field solution (13). The FE algorithm is fully versatile in this regard, including domain multiple-connectivity. For a diffused HVAC supply stream into a 2D chamber, with "floating" center divider, a *totally* symmetric velocity field as obtained with uniform exhaust port pressure levels, Fig. 2a. Conversely, Fig. 2b

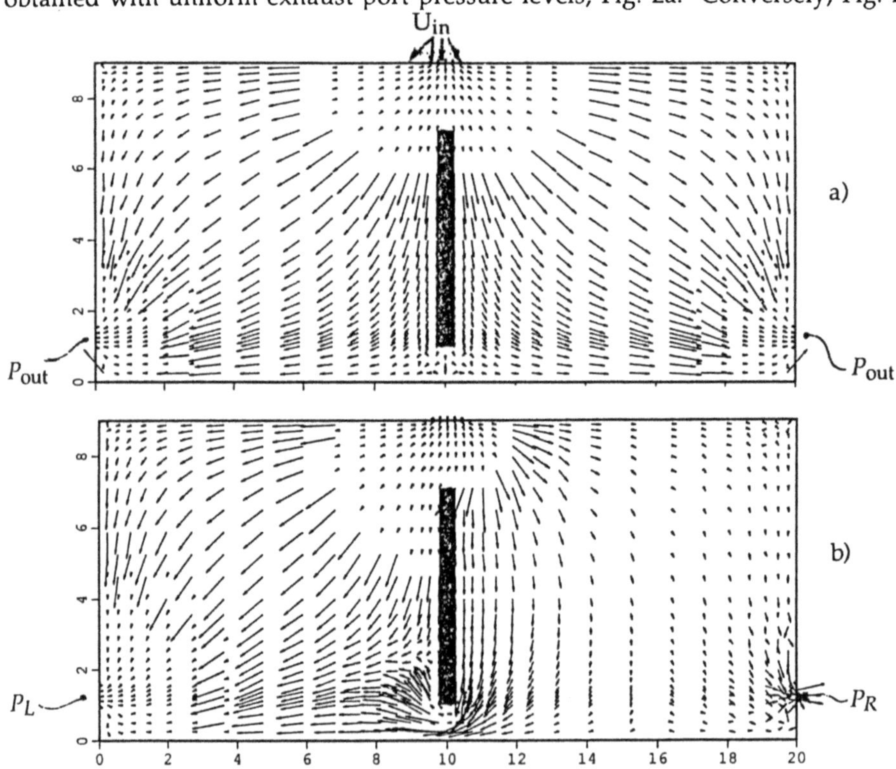

Figure 2. Flowfield in multiply-connected domain with dual exits, $Re/=10^3$, U_{in}=10fps, a) equal exit pressures, b) $P_R > P_L$ differential exit pressure; arrow tails for $u/U_{in}>0.1$ are suppressed for clarity.

shows the non-symmetry induced by an exhaust port pressure differential of about 1 cm water. A further ~50% differential increase renders the right exhaust an inlet, illustrating flow total redirection as induced by minor exhaust pressure level changes.

Flow obstruction: The HVAC supply outflow, in the AC mode, is typically directed at the passenger's torso. The cabin computational domain is multply-connected, hence the flow bifurcates about a "stagnation plane." For a 2*D* model geometry consisting of three stagnation planes in symmetrical disposition, Fig. 3a, and for symmetric inflow, the velocity field must bifurcate symmetrically with no flow passing through the interior regions. For Re^f=500, Fig. 3b confirms the resultant symmetric CFD solution, as obtained using flow tangency boundary conditions on the planes. Heating the planes, i.e., a "thermal person," would destroy this symmetry via buoyancy effects.

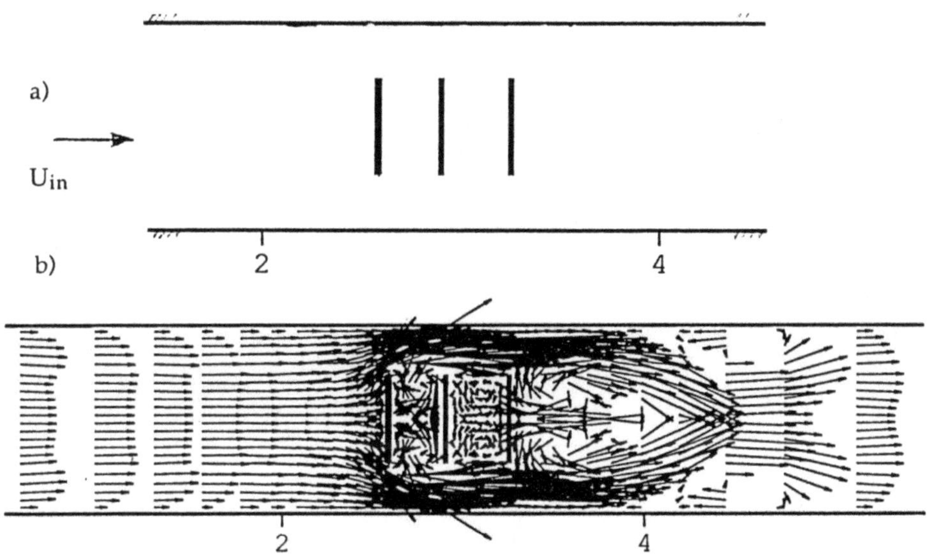

Figure 3. Stagnation flow in multiply-connected domain, Re^f=500, U_{in}=1fps, a) geometry for three dividers, b) velocity vector field, symmetric inflow.

Buoyancy driven flow: Natural convection can play a significant roll in a cabin compartment hot soak situation with radiation loading of closed windows. The benchmark problem is the hot/cold wall "thermal cavity" without flow outlets. The lack of a characteristic velocity results in Gr/Re^2 in (2) becoming

RaPr, and transition to turbulent flow occurs for Rayleigh number (Ra) exceeding about 10^7. Further, no suitable pressure or constraint field Dirichlet data reference exists, which restricts available linear algebra procedures. A mesh refinement study led to a $(101)^2$ nodal mesh, non-uniform for wall layer resolution, with local mesh aspect ratios exceeding 50:1, Fig. 4a. Steady state solutions were obtainable to Ra=10^6; the isotherm distributions were as expected [5], c.f., Fig. 4c, and an excellent pressure solution was obtained, Fig. 4d.

a) b)

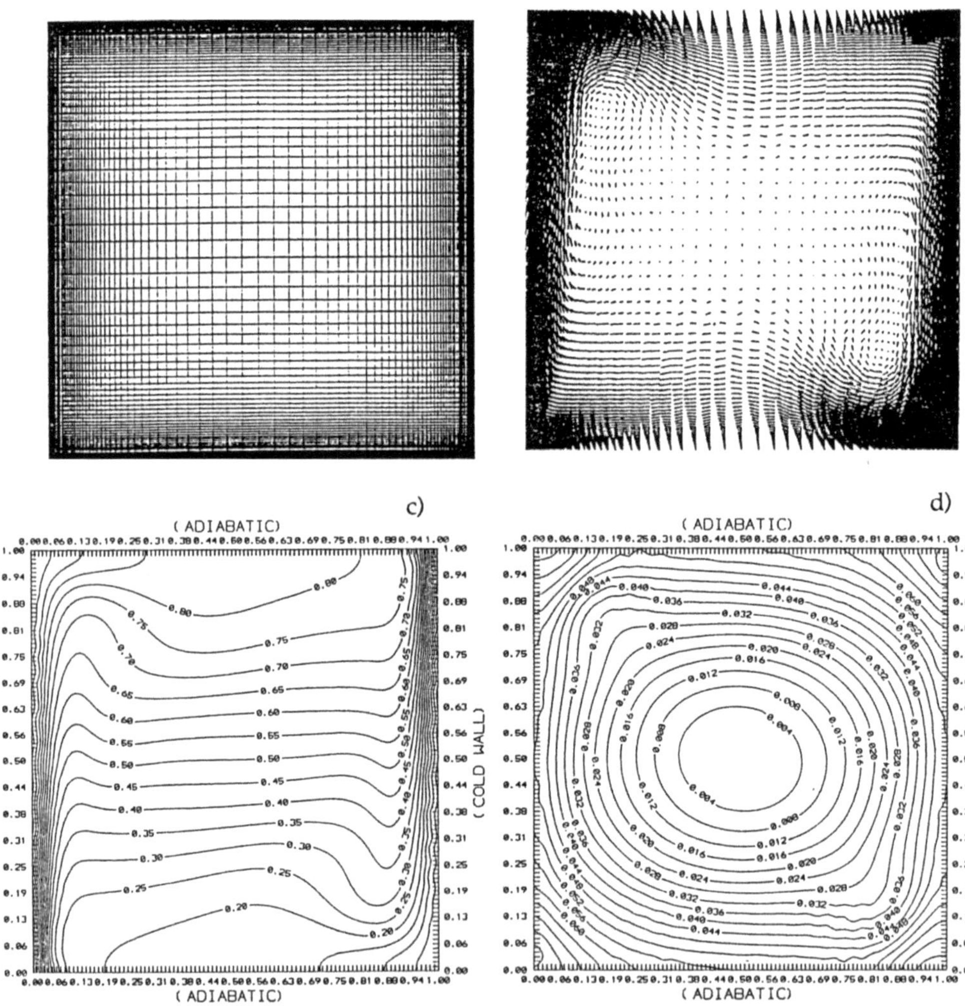

Figure 4. Natural convection in a square cavity, $\Omega^h=100^2$, Ra=10^6,
a) nonuniform mesh, b) steady velocity field, c) isotherms, d) isobars.

SUMMARY AND CONCLUSIONS

An inexact, continuity-constraint iterative CFD theory for the incompressible-thermal Navier-Stokes PDE system is analytically established. Implementation into a computable form (code) combined a FE semi-discretization of a Galerkin weak statement with an implicit time-integration procedure. The resultant CFD algorithm/code employs a non-staggered mesh, and is intrinsically free of artificial numerical diffusion. Benchmark test results document robust handling of issues critical to simulation of passenger compartment HVAC flowfields. Development of the ARINS algorithm "template," for COMCO's *AKCESS.*™ UNIX production shell, is proceeding.

ACKNOWLEDGMENTS

The authors wish to acknowledge the considerable contributions of Messrs. E. G. Schaub and P. T. Williams in design and execution of the computational experiments.

REFERENCES

1. M. R. Heller (ed), Springer-Verlag, Berlin, FRG, 1989.
2. A. J. Baker and D. W. Pepper, McGraw-Hill, New York, NY, 1991.
3. A. J. Baker, Hemisphere, Washington, DC, 1983.
4. B. F. Armaly, et. al., J. Flu. Mech., V. 127, 1983.
5. G. de Vahl Davis, Int. J. Num. Mtd. Fluids, V. 3, 1983.

A Study of Aerodynamic Characteristics of Detailed Car Configurations Using a Finite Difference Analysis of Airflows

Ryutaro Himeno
Central Engineering Laboratories
Katsuro Fujitani and Sanae Satoh
Product Development Systems Department
Nissan Motor Co., Ltd.
1, Natsushima-cho, Yokosuka 237, Japan

Summary

Both external and internal flows of cars were simulated. A third-order upwind-difference scheme was used in these simulations. Computational grids were generated by a multi-block transformation and a trans-finite method. Engine compartments were modeled in grid systems. They had engine blocks, transmissions, air-cleaners, and suspension arms. Heat-exchangers were simulated as pressure losses. In the first case, influences of a license plate and bypass-flows on the cooling airflow were investigated. Simulated results were compared with experimental results using a production car. In another case, cooling airflows of a high-speed research car were studied. Computed results told that the amount of cooling airflow through inter-coolers and oil-cooler were not sufficient in the first design stage. This was also found in the test of an actual car. We concluded that the present method is quite useful to predict air flow rates through heat exchangers built in a new car in the early design stage.

Introduction

Computational Fluid Dynamics (CFD) has developed greatly due to the progress of supercomputers and computational schemes. Various 3-dimensional analyses of the airflow around an automobile are carried out nowadays[1-3]. The present authors also have simulated flows related to automobiles and reported that a third-order upwind-difference scheme has enough accuracy to predict their aerodynamic characteristics[4, 5]. It has become possible to predict a drag coefficient of a production car with its error less than 2 percents[5]. Simulation of cooling flow is another important application of CFD. The shape of inlets is important in both styling and engineering points of view. It is quite useful to be able to predict the amount of cooling airflow through heat exchangers on early design stage. For this analysis, it is essential to calculate both external and internal flows simultaneously.

There are several papers which simulate engine cooling flows of cars[6-8]. However, almost all of them are using rectangular grid systems which are not suitable to calculate external flows along smooth surfaces. Now we report on the application of a third-order-upwind scheme combined with a curvilinear coordinate system to the prediction of cooling airflows of cars.

Basic Equations

The airflow around automobiles is incompressible. The continuity and Navier-Stokes equations can be described as follows with a coordinate system fixed on the vehicle moving with velocity v_0 in a stationary fluid.

*Equation of continuity

$$\nabla \cdot v = 0 \tag{1}$$

*Navier-Stokes equation

$$\frac{\partial v}{\partial t} + \{ (v - v_0) \cdot \nabla \} v = -\nabla p + \frac{1}{Re} \Delta v \tag{2}$$

The Poisson equation with respect to pressure of high Reynolds number flows is as follows,

$$\Delta p = -\nabla \cdot \{ (v - v_0) \cdot \nabla \} v - \frac{\partial D}{\partial t} , \tag{3}$$

where $D = \nabla \cdot v$. Then the second term of the right hand side of this equation is expressed by a finite difference approximation. Although D should be 0 due to the equation of continuity, D^n is not omitted so that any error in satisfying the equation of continuity at the present time step n can be compensated. Finally, pressure equation to be solved becomes as follows:

$$\Delta p = -\nabla \cdot \{ (v - v_0) \cdot \nabla \} v + \frac{D^n}{\Delta t} , \tag{4}$$

where Δt is a time increment. The equation of continuity is not directly satisfied but implicitly satisfied. This derivation is based on a well-known MAC method[9], which may cause the error in the flow quantity in case of internal flows but may not cause significant errors in case of external flows. Next, eqs. (2) and (4) are transformed into a generalized coordinate system and discretized. A second-order implicit Euler method is used for time derivatives. Convective terms are discretized using a third-order upwind-difference scheme[10]. A second-order central-difference scheme is used for other spatial derivative terms.

Computational Grid System

A multi-block-transformation is used to generate grid systems. This method is more suitable for the analysis of the flow around objects with complicated geometry than a single-block approach[11].

For this grid generation, the grid distribution on the boundary surfaces of the computational domain is determined first. The surface data on the vehicle are given from CAD data. The grid in each block is generated by a transfinite interpolation[12] using the grid distributions on the boundaries. At the next step, all blocks are combined into one block. Finally, it is modified to be smoother and orthogonal to the surfaces. These procedures are described more in detail in the reference 13.

Figure 1 and 2 show a grid system for a three-box sedan. Lower suspension arms, a power train, a radiator and its support are modeled in the grid system(see figure 2). Figure 3 shows surface grid distribution of a high-speed research car. The dimensions of the computational space in both cases are 2 times the body length upstream, 10 times downstream, 5 times the width sideward and 5 times the height upward, respectively. The minimal spacing of grid points is 1/10,000 of the body height. Those grid systems have 180x131x80 grid points and 221x67x81, respectively. It took about 6 weeks for one person to generate a grid system for the three-box sedan. However, the modification between two cases took 1 week. Grid generation of the research car took about almost 3 months. A symmetric condition is assumed in the calculation of flows around the research car. However, the power-train for the three box sedan is not symmetric (see figure 3) and symmetric condition is not applied in this case. Heat exchangers are modeled as pressure loses which are proportional to the dynamic energy of flows.

Computer System

The computer hardware used for this study are shown in figure 4. A mini-supercomputer

Convex C240 is used for the test of grid systems, the computation of particle tracing and also a file server. Graphic work stations are used for grid generation and the display of computed results. Video tape recorders(VTR) are connected to graphic workstations through a controller for taking pictures frame by frame to make animation films. Computations of flows were carried out on a CRAY YMP8/664.

Series of unsteady flow fields of all grid points were output into files every hundred-time steps. Streak lines are calculated from those stored data of unsteady flow fields on C240. In these calculations, a linear interpolation is used for unsteady flow fields and a second-order Euler integration was used for time-integration of particle traces.
All programs for flow visualization are written in C-language. The program for particle tracing is vectorized and parallelized on the C240.

Results and Discussion

Reynolds number based on the hight of the car is $2x10^6$ in every cases. In the comparison between calculations and experiments, the vehicle configurations and Reynolds number are the same in both cases. Flow rates through heat exchangers were measured in a full-scale wind tunnel using actual cars.

In all calculations, cars are moving with velocity v_0 in stationary fluids, and the boundary condition of the ground is set as stationary with velocity 0 in order to simulate the real driving situation.

Three-Box Sedan

The amount of cooling airflow through a radiator mounted on an actual front-wheel-drive three-box sedan were calculated and compared with experimental data in two cases. This computation took about 30 hours on CRAY YMP8/664 in each cases. Figure 5 shows its model. The difference in exterior between two cases is whether it has a license plate or not. The car with a license plate has holes on a radiator support as shown in figure 6 in order to simulate bypass flows. These holes are for headlights and their area are set as much as a half of total holes or space passing through the radiator support except for the radiator.

Figure 7 and 8 show instantaneous streamlines. Every pair of pictures have same start points for streamlines. It is obvious that the license plate disturbs a flow coming into the engine compartment. However, it seems that bypass holes do not have a significant influence on the flow field.

Time-averaged airflow rates are shown in table 1 compared with experimental measurements. In experiments, pressure differences between both sides of the radiators on 40 points were measured. Then, their velocities are look up from a table which tells a pressure loss on a certain velocity. A flow rate is defined here as averaged velocity through a heat exchanger against a mean flow. Experiments were performed using same configuration as grid systems in each cases. Calculated flow rate in the case of the car without the license plate agrees excellent with experiments. However, there is a considerable discrepancy in the case of the car with the license plate. This may be caused by bypass holes. Tested car had distributed bypass holes around its radiator. The area of bypass hole may be under-estimated. The other possibility may be the effect of the position of holes, as a hole near the radiator should have more effect than far one.

High-Speed Research Car

The computation of a flow around the high-speed research car took about 40 hours on CRAY YMP8/664. Figure 9 shows streak lines around the research car. A flow passing through a front radiator can be seen in this figure. There are also inlets on the front end for air ducts of break-cooling in the grid system.

Table 2 shows comparison of flow rates of heat exchangers between simulation and experiment. Their agreement show quite well. From engineering point of view, these flow rates of oil coolers and inter-coolers are not good enough. Pressure distributions in figure 10 were extremely useful for engineers to modify these cooling system.

Conclusion

Airflow rates of heat exchangers were predicted using a third-order upwind-difference scheme combined with a curvilinear coordinate system. We found that the present method can simulate flows well and that the accuracy is reasonable. A multi-block transformation is powerful enough to generate complicated grid systems for these computations. Computation time is so long that this type of calculation is not practical now. In near future, an improvement of supercomputers will make it practical.

References

[1] Shaw, C.T. and Simcox, S., The Numerical Prediction of the Flow Around a Simplified Vehicle Geometry, Proceedings of the Second International Conference on Supercomputing Applications in the Automotive industry, Seville, Spain, October 1988, A Cray Research, Inc. book, pp.219-232, 1988.

[2] Hutchings, B.J. and Pien, W.S., Computation of Three Dimensional Vehicle Aerodynamics Using FLUENT/BFC, Proceedings of the Second International Conference on Supercomputing Applications in the Automotive industry, Seville, Spain, October 1988, A Cray Research, Inc. book, pp.233-256, 1988.

[3] Kawaguchi, K., Hashiguchi, M., Yamasaki, R. and Kuwahara, K., Computational Study of the Aerodynamic Behavior of a Three-Dimensional Car Configuration, SAE paper, 890598, 1989.

[4 Fujitani, K., Himeno, R. and Takagi, M., Unsteady 3-D Calculation of Flow around a Car and its Visualisation, Automotive Simulation: Proceedings of Second European Cars/Trucks Symposium, Schliesee, FRG, May 1989, Moshe R. Heller (Editor), Springer-Verlag Berlin, Heidelberg, 1989.

[5] Himeno, R., Takagi, M., Fujitani, K. and Tanaka, H., Numerical Analysis of the Airflow around Automobiles Using Multi-block Structured Grids, SAE Paper 900319, 1990.

[6] Kuriyama, T., Numerical Simulation on Three Dimensional Flow and Heat Transfer in the Engine Compartment Using "STREAM", Proceedings of the Second International Conference on Supercomputing Applications in the Automotive industry, Seville, Spain, October 1988, A Cray Research, Inc. book, pp.283-292, 1988.

[7] Shibata, S., Hosaka, S., Fujitani, K. and Himeno, R., A Numerical Analysis Method for Optimizing Intercooler Design in the Vehicle Development Process, SAE Paper 900080, 1990.

[8] Aoki, K., Numerical Simulation of Three Dimensional Engine Compartment Air Flow in FWD Vehicles, SAE Paper 900086, 1990.

[9] Roache, P. J., Computational Fluid Dynamics, Hermora Publishers, 1976.

[10] Kawamura, and Kuwahara, K., Computation of High Reynolds Number Flow around a Circular Cylinder with Surface Roughness, AIAA paper, AIAA-84-0340, 1984.

[11] Sawada, K. and Takanashi, S., A Numerical Investigation on Wing/Nacelle Interfaces of USB Configuration, AIAA paper, AIAA-87-0455, 1987.

[12] Tompson, J. F., Warsi, Z. A. U. and Mastin, C. W. Numerical Grid Generation : Foundations and Applications, North-Holland, 1985.

[13] Fujitani, K. and Himeno, R., A CAD data oriented grid generation system and its application to automotive aerodynamics, Numerical Grid Generation in Computational Fluid Dynamics and Related Fields, Proceedings of the Third International Conference on Numerical Grid Generation in Computational Fluid Dynamics and Related Fields, Barcelona, Spain, 1991, A.S.-Arcilla, J.Hauser, P.R.Eiseman and J.F.Thompson (Editors), North-Holland, Amsterdam,1991.

Table 1. Comparison of flow rates of three-box sedan with experiments

	Simulation	Experiment	Difference
Without License Plate	15.1%	14.2%	0.9%
With License Plate	9.0%	6.9%	2.1%

Table 2. Comparison of flow rates of a research car with experiments

	Simulation	Experiment	Difference
Front Radiator	20.3%	17.0%	3.3%
Side Inter-Cooler	3.3%	5.0%	-1.7%
Rear Oil-Cooler	2.1%	1.0%	1.1%

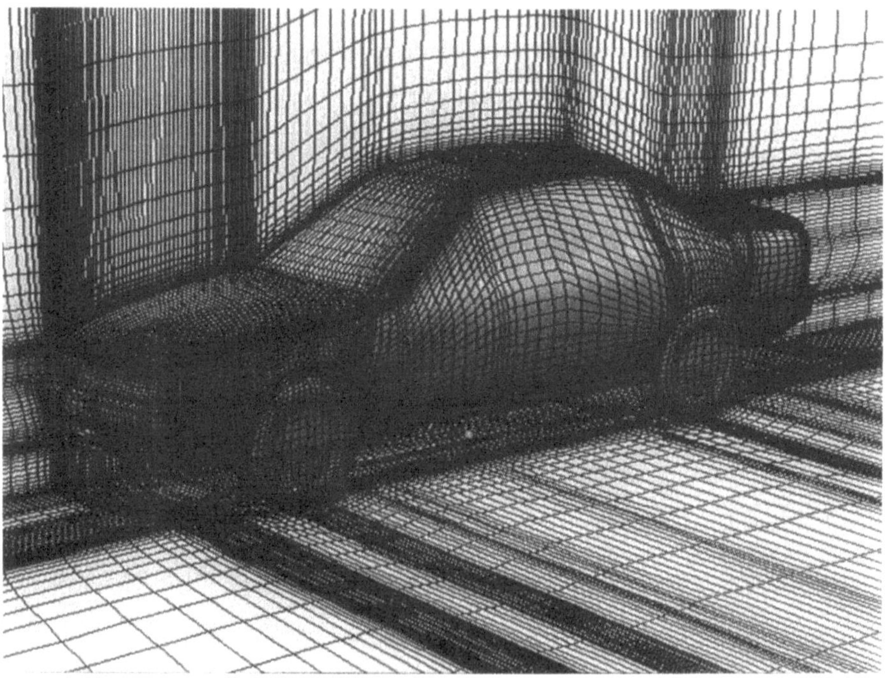

Figure 1. Grid system of a three-box sedan. (180x131x80 grid points)

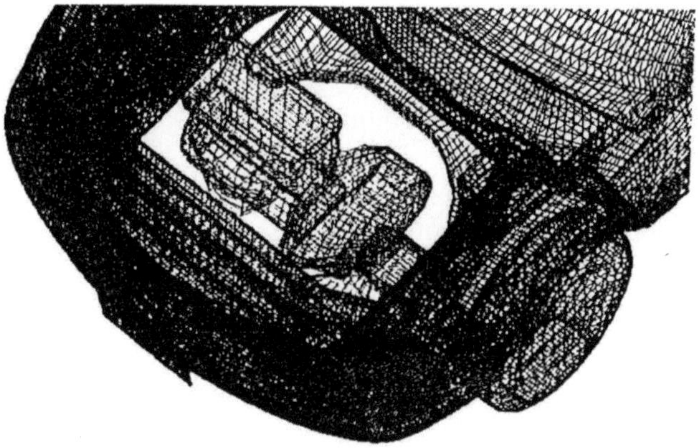

Figure 2. Surface grid distributions in the engine compartment of a three-box sedan.

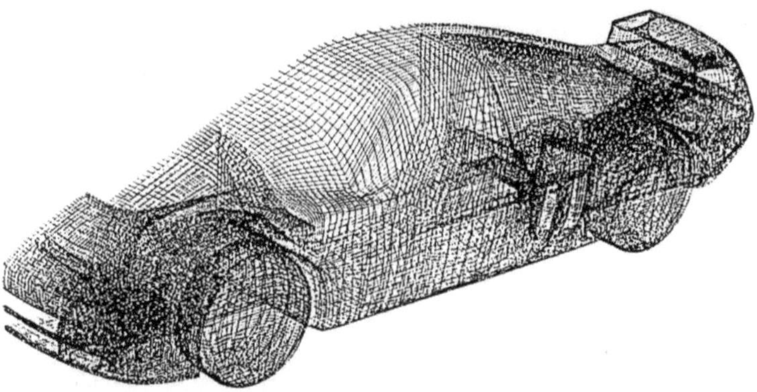

Figure 3. Surface grid distributions of a high-speed research car. (221x67x81)

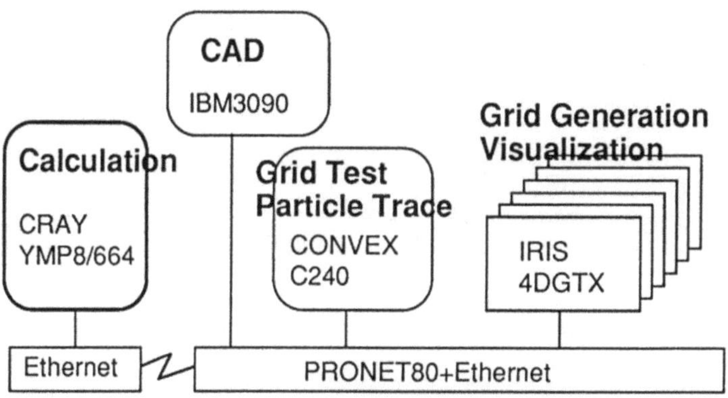

Figure 4. Computer system.

Figure 5. Shaded image of surface grids of a three-box car without a license plate.

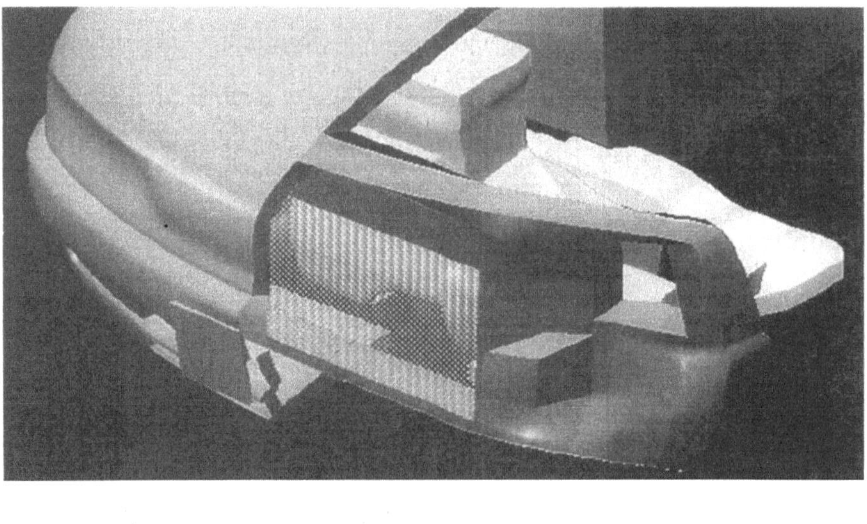

Figure 6. Shaded image of a car with a lcense plate and a radiator support with a bypass hole.

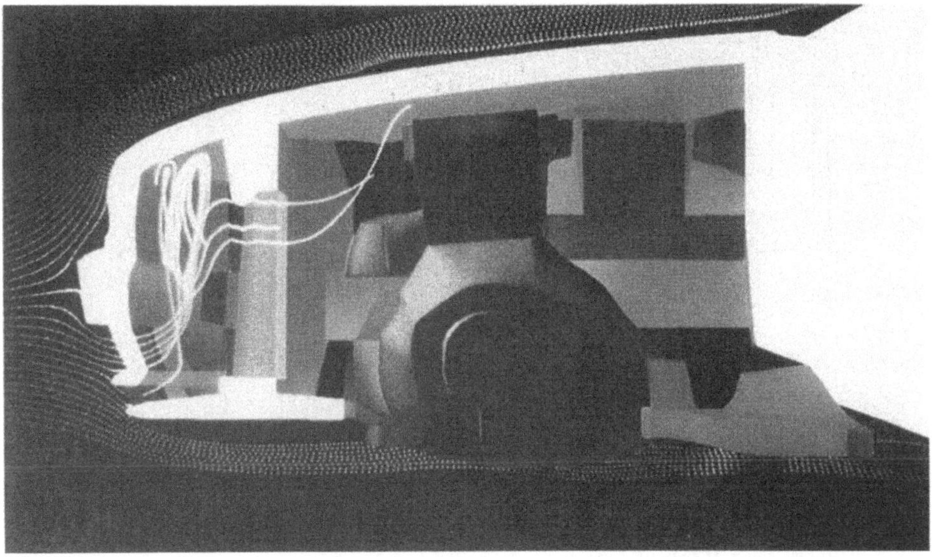

a) Without a license plate.

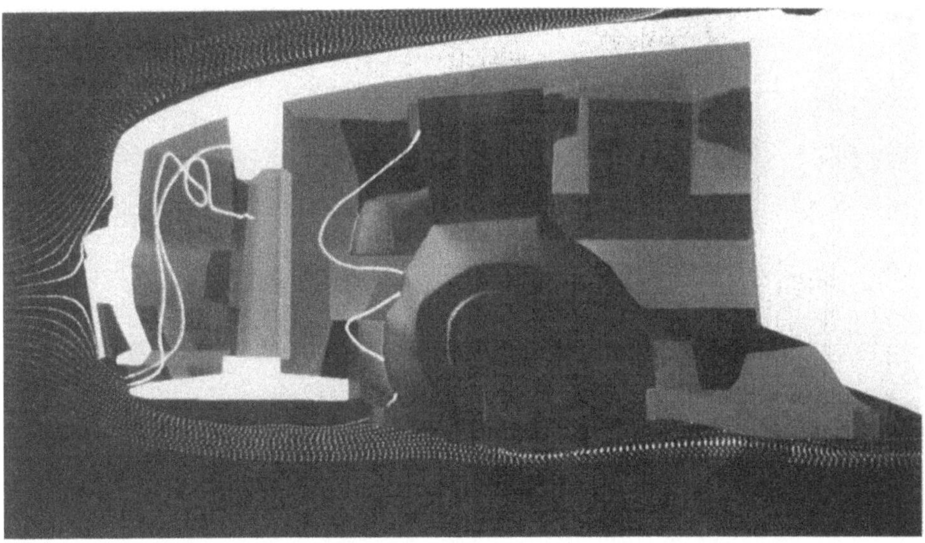

b) With a license plate.

Figure 7. Comparison of flows between a car without a license plate and one without. (Sectional view)

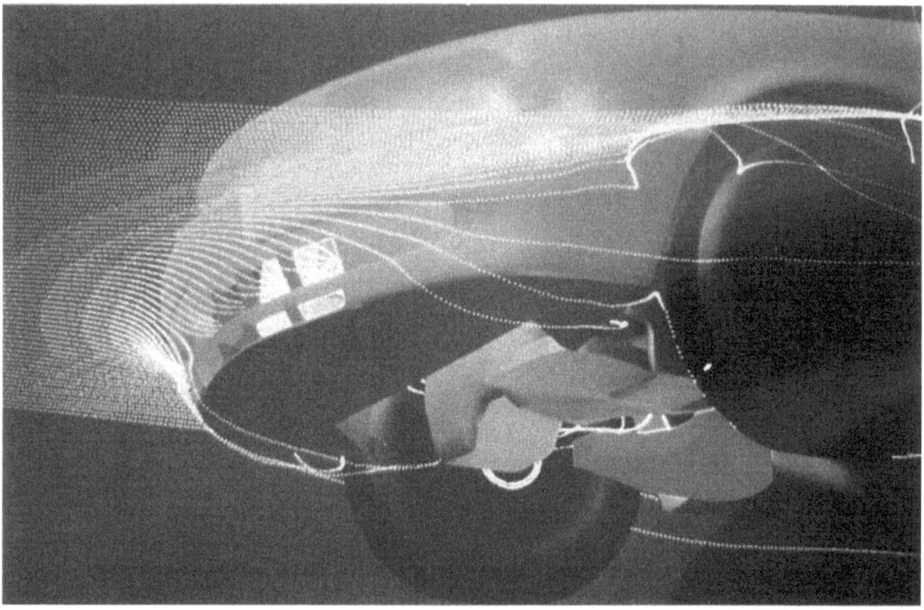

a) Without a license plate.

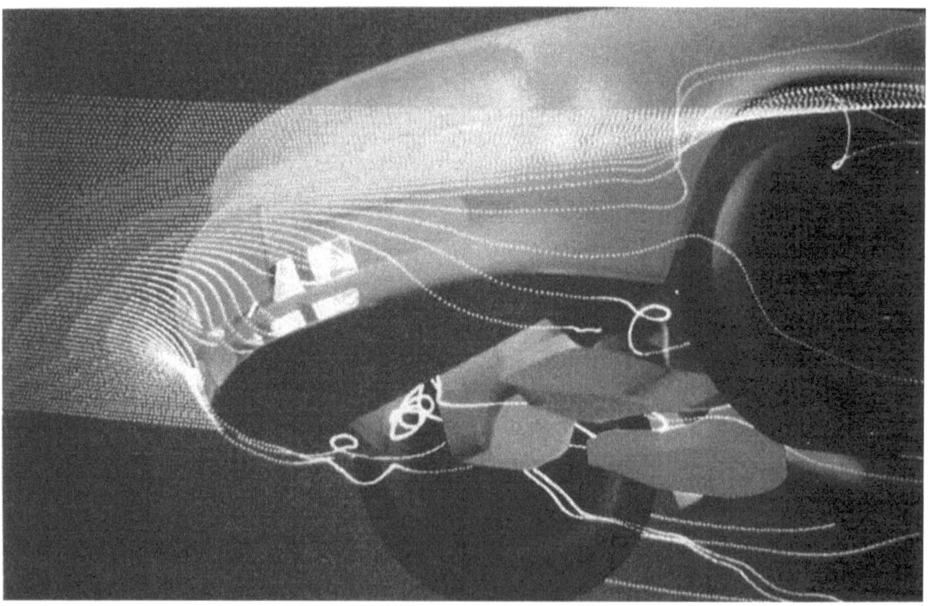

b) With a license plate.

Figure 8. Comparison of flows between a car without a license plate and one without. (View from ground)

Figure 9. Streak lines around a high-speed research car.

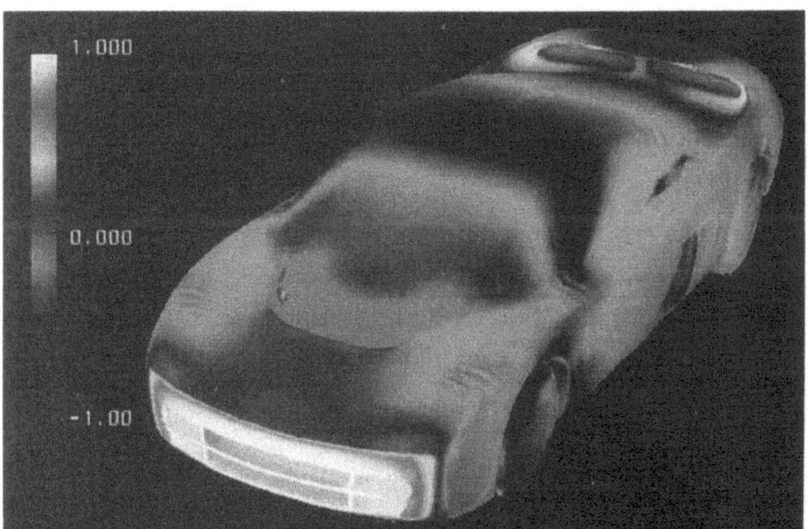

Figure 10. Pressure distributions on body surfaces of a research car.

Ventilation Duct Design by CFD-Technology

Dietmar Fischer, Ford Werke AG Koeln

Abstract

A general objective for designing ventilation ducts is to minimize the pressure drop between inlet and outlet. Moreover, the velocity profile at the outlet has to comply with special requirements as well. Package restrictions in modern passenger cars lead to complex paths of these ducts, which make a design optimization difficult.
This paper presents a possible design process for ventilation ducts with Computational Fluid Dynamics (CFD) and the validation of this method.
The CFD code STAR-CD was used to simulate the steady state, incompressible and turbulent 3D-flow.
Various design alternatives were analysed with this method. The pressure drop could be reduced by about 60 % and the flow profile at the outlet was significantly improved.
For the cases where experimental data is available a good correlation with the test results has been found.
In conclusion the CFD method is a good tool to improve the design of ventilation ducts.

Contents

1. Objectives

* Optimize the current design of the ventilation duct regarding the pressure drop and the outlet velocity profile.

* Demonstrate the ability of Computational Fluid Dynamics (CFD) to predict the pressure drop in ventilation ducts.

* Compare the numerical results with measurements.

* Show the benefits of the CFD based design method compared to the current test based method.

2. Background

CFD is a CAE method, which recently became more important due to the increasing performance of the computers. CFD can address a number of development tasks within automotive engineering.
Here the simulation of the flow in ventilation ducts was performed with CFD to improve the upfront engineering of these ducts.
One important task of these simulations is to predict the pressure drop caused by the ventilation ducts, because this pressure drop influences directly the power required for the fan. High pressure drops in ventilation ducts or other components of the climate control system require highly powered fans compared to optimized systems.
In addition the outlet velocity profile of the ducts has to be as uniform as possible to ensure the function of the movable vanes at the outlet nozzle. High velocities and turbulence quantities have to be avoided in order to reduce the noise radiation of the flow inside the ducts. All these facts lead to the necessity to optimize these components.
Due to the package requirements of modern passenger cars, the design of ventilation ducts is nowadays more complicated than it was in the past. The current practice in designing ventilation ducts is mainly based on experiments which are very time consuming, because the hardware has to be built first. This practice led in the past to long development phases in conjunction with expensive changes of the manufacturing tools.
Furthermore usually no detailed information about the flow inside the ducts is available from the experiments. A very high amount of work is necessary to obtain this information from experiments.
But this detailed flow information could help the engineer to optimize his design and to understand the fluid behaviour.
Therefore it was tried here to improve the current practice by introducing a CFD based method where the optimization can be performed on a computer without building hardware models of the ducts.
The investigation was performed on the codriver side ventilation duct of a passenger car placed between the heat exchanger and the codriver side ventilation nozzle.
As shown in /1/, /2/ and /5/ the accuracy achieved by the current CFD method is not extremely high. Differences between experiments and calculations have to be expected.

3. Analyses

3.1 Type of Simulation

The analyses were performed with the commercial CFD code STAR-CD 1.813 described in /3/.
The problem was assumed as steady state, incompressible and turbulent 3D-flow. Buoyancy effects were not included. The fluid was air at a temperature of 20 degrees centigrate.

* Steady state means that there are no significant transient effects in this problem except the turbulence. The transient turbulent fluctuations of the velocity were simulated by use of the k - ϵ - model of turbulence which models here a time independent turbulent viscosity. The standard coefficients were used for the k - ϵ - model /3/.

* Incompressible means that the variation of the fluid density in the whole flowfield is small. This condition is fullfilled in this problem because the fluid has a uniform temperatur in the whole flow field and the Mach-number is very low (< 0.03).

* The flow is turbulent because the Reynolds-number is about 30000, which is much higher than the threshold for laminar flow in pipes (2300).

* A 3D-simulation was necessary due to the complex shape of the ducts.

* The buoyancy effects were not included, because the density of the fluid is nearly constant and the gravitation has small influence only.

The whole flowfield was modelled by a maximum of 60600 control volumes (cells). For these models the following six equations were solved: The X-momentum, Y-momentum, Z-momentum, Continuity, Turbulent-Kinetic-Energy and Turbulent-Energy-Dissipation .
The first order Upwind Differencing Scheme (UDS) was used. It is worth to mention that this scheme produces numerical errors, which sometimes are called false or numerical diffusion /4/. On the other hand the UDS convergence is quite stable and does not tend to produce non-physical overshoots like higher order schemes tend to do /6/.

3.2 Mesh generation

To generate the simulation model as efficiently as possible a program was developed which generates a model from CAD data semi-automatically by dragging a section discretization along an unlimited number of drag-lines as shown in Att. 3.2 .
This method allows the user to change the shape of a ventilation duct easily on a CAD system taking all package requirements into account and to build a new simulation model in a very short time.
This is certainly very helpful for supporting the design process of ventilation ducts efficiently.

3.3 Mathematical Models

Four mathematical models were generated on the basis of CAD data:

Model 1 :48000 cells	model of the base version of the ventilation duct without the distribution box at the inlet and without the nozzle at the outlet. (For comparison with experiments only)
Model 2 :60600 cells	model of the base version of the ventilation duct with the distribution box at the inlet and with the nozzle at the outlet.
Model 3 :60600 cells	improved model of the ventilation duct with a smoothed distribution box at the inlet and with the base version of the nozzle at the outlet.
Model 4 :60600 cells	Complete smoothed and changed design including the distribution box at the inlet and the nozzle at the outlet. A turning vane was included in the nozzle region.

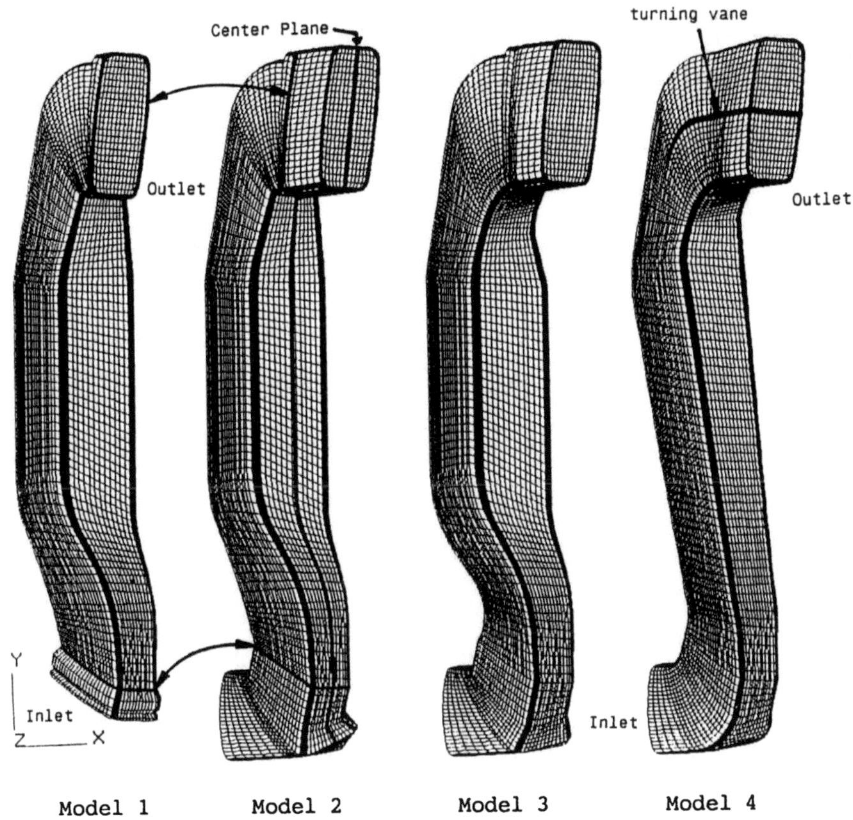

Fig. 3.3.1 Models 1-4 shown in a Plan View . Center Plane Example used for the following Result Display.

3.4 Boundary Conditions

- The flow through the inlet cross section of each model was defined to have a constant velocity of 6.78 m/s. This is equivalent to a flow rate of 36.0 l/s, which is an usual flow rate for this ventilation duct under working conditions.
  Since the pressure drop ratio of several designs is similar at all measured flowrates shown in Att. 3.4.1, it is usually sufficient to investigate a duct at one flowrate only. Due to a lack of experimental data the inlet turbulence quantities were defined to be zero.
- The outlet cross section was assumed to have a constant zero pressure distribution. The turbulence quantities at the outlet were defined to be zero for backflow like it was defined at the inlet.
- The walls of the duct were simulated by the logarithmic LAW OF THE WALL as hydraulically smooth.

3.5 Solution Statistics

- About 500 iterations were necessary for a solution to converge.
- The CPU time used per iteration for model 2, 3 and 4 was about 20 seconds on the Cray-YMP and about 150 seconds on a workstation DEC 5000/200.
- The memory required for model 2, 3 and 4 was about 7.5 mega words on the CRAY-YMP and about 30 mega bytes on a DEC 5000/200.

3.6 Results

3.6.1 Resultant Velocity Distributions

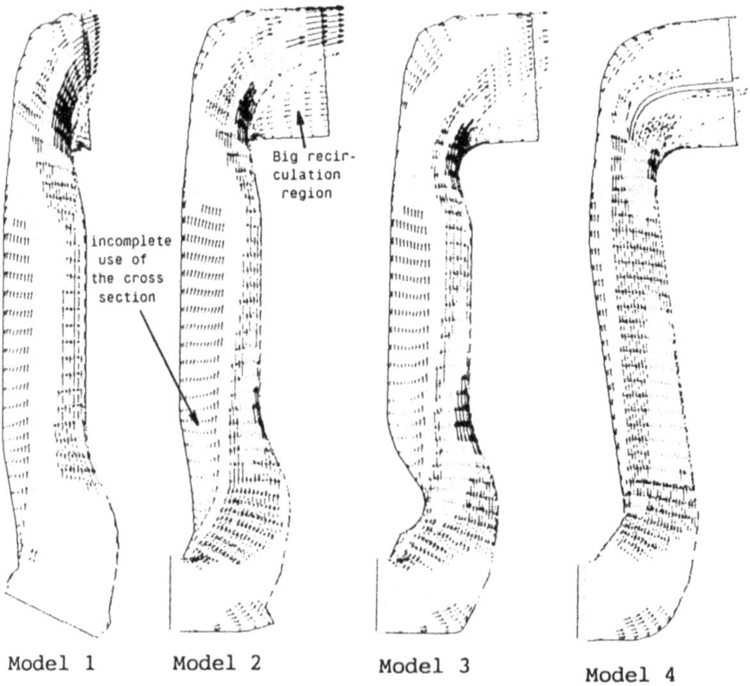

Fig. 3.6.1.1 Velocity Distributions in the Center Plane

Fig. 3.6.1.1 shows the velocity distribution in the center plane of all calculated ducts.
The basic model 2 is not properly designed at the inlet as well as at the outlet. This leads to an incomplete use of the cross section due to the inlet design and a big recirculation region due to the design of the bend at the outlet. Also large local velocities, which are shown as dark areas in Fig. 3.6.1.1, may lead to high acoustic radiation. All these shortcomings were partially resolved by model 3 and almost completely resolved by model 4.
The objectives of the optimization process are to achieve a uniform velocity distribution through the outlet nozzle and to minimize the pressure drop induced by the ventilation duct.

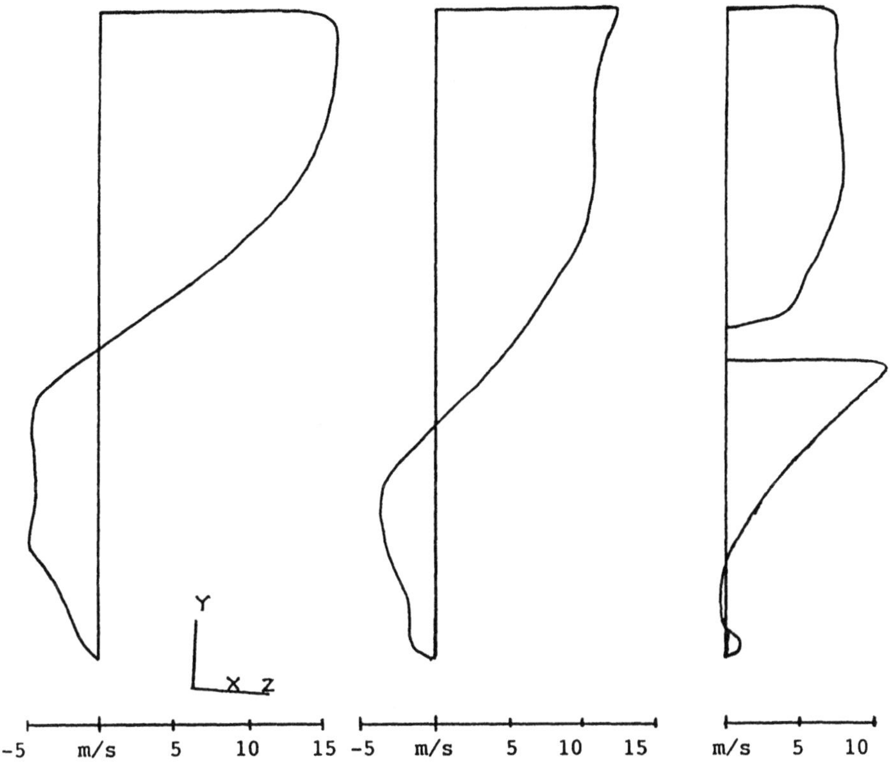

(- indicates backflow)

Model 2 Model 3 Model 4

Fig. 3.6.1.2 Velocity Distribution in the Center Plane of the Outlet

Fig. 3.6.1.2 shows the velocity distribution in the center plane at the outlet of model 2 to 4 and the achieved improvement by the various duct designs.
The most uniform velocity distribution was achieved with model 4. A further approach to the objective requires the implementation of more turning vanes and/or the increase of the outer dimensions of the duct, which is restricted by package requirements. Therefore the optimization process was stopped here.

3.6.2 Resultant Pressure Drop

As mentioned before one objective of the optimization process is to minimize the pressure drop induced by the ventilation duct.
For comparison purposes with experiments the pressure drop definition was chosen to be in line with the data obtained from the test-equipment shown in Att. 4.1.1 .
Since the whole test equipment was not modelled in the analyses, a pressure drop of one stagnation pressure due to the mixing of the outlet jet inside the test-chamber was assumed. The pressure drop for the smoothed inlet was assumed to be zero.
Then the averaged total pressure at the inlet represents the pressure drop, because the pressure distribution at the outlet was defined to be constant zero by means of the boundary conditions described in 3.4.

It should be remembered that due to the reasons described in 3.3 only model 2, 3 and 4 can be compared.

Model No.	Calculated Averaged Total Pressure at the Inlet [N/m2]
1	221
2	240
3	170
4	94

Tab. 3.6.2.1 Calculated Average Total Pressure of the Inlet Cross Section for Model 1 to 4.

Tab. 3.6.2.1 shows the achieved reduction of the pressure drop. The reduction of the pressure drop from model 2 to model 4 was about 60 %. This is equivalent to a reduction of the necessary performance of the fan if the analysed ventilation duct was the only component of the ventilation system.

4. Validation

4.1 General Remarks

The use of CFD for automotive development purposes is quite new and is therefore not validated in all engineering applications.
Also the current state of the art in CFD has several shortcomings specially in modelling the turbulence and in the simulation of the near wall region. This requires validations by using experimental data.

The validation for the ventilation duct investigation described here was done in accordance with the objective to achieve a uniform velocity distribution at the outlet and to minimize the pressure drop.

Experimental data of the pressure drop is available for model 1, 2 and 3. Experimental data of the velocity distribution at the outlet is available for model 2 and 3. Experimental data of noise amplitudes is currently not available.
The inlets of the tested ducts were smoothed to achieve a uniform velocity distribution over the inlet cross section like it was defined by the boundary conditions of the analyses described in 3.4 .

The available experimental velocity data at the outlet was obtained from fan anemometry measurements. As shown in Att. 4.1.2 this method leads to a high blockage of the outlet cross section due to the anemometer housing which of course influences the flow in the duct. This has to be taken into account, if the results of the analyses and the tests are compared.

4.2 Validation of the calculated pressure drop

Model No.	Pressure Drop [N/m2]		Difference [%]
	Analysis	Test	
1	221	250	11.6
2	240	270	11.1
3	170	155	9.7
4	94	-	-

Tab. 4.2.1 Comparision between calculated and measured pressure drop

The calculated results show the same trend as the test results. This means that a calculated reduction of the pressure drop could be identified during the tests as well.
Taking the test data as reference the comparision of Tab. 4.2.1 indicates a maximum difference between calculation and test of about 12 % for the overall pressure drop in these ventilation ducts.
For engineering purposes this is a good correlation. As described in 3.6.2 some assumptions had to be made to compare the tested data with the calculated one. Therefore it is possible that the remaining difference is due to shortcomings of the simulation as well as to the assumptions made in 3.6.2 .

4.3 Validation of the calculated velocity distribution at the outlet

As shown in Fig. 4.3.1 the calculated and measured velocities show the same trend. Taking the maximum velocity as a reference, the largest difference is about 20 %.
As mentioned before, the fan anemometry measurement method is not a very accurate method. On the other hand this method is easy to handle and is often used in the engineering practice.
Therefore the differences between experiment and calculation may be due to the measurement method as well as to the calculation.

Recent hot-wire anemometry measurements on a similar duct led to the result, that the fan-anemometry method tends to underpredict the velocity values by about 15% compared to the values obtained by the hot-wire anemometry method. This fact would reduce the differences between test and calculation.

4.4 Mesh Influence of the Results

The base version 2 of the ventilation duct was selected to investigate the influence of the discretisation on the calculated results.
This model was generated once with 60600 cells as described in 3.3 and once with 15150 cells.

Fig. 4.4.1 shows the two versions of the model which were analysed. The discretisation of the cross section was changed from 600 of model 2 to 150 of model 2c. The discretisation of the longitudinal direction of the duct was 101 cells in both cases.

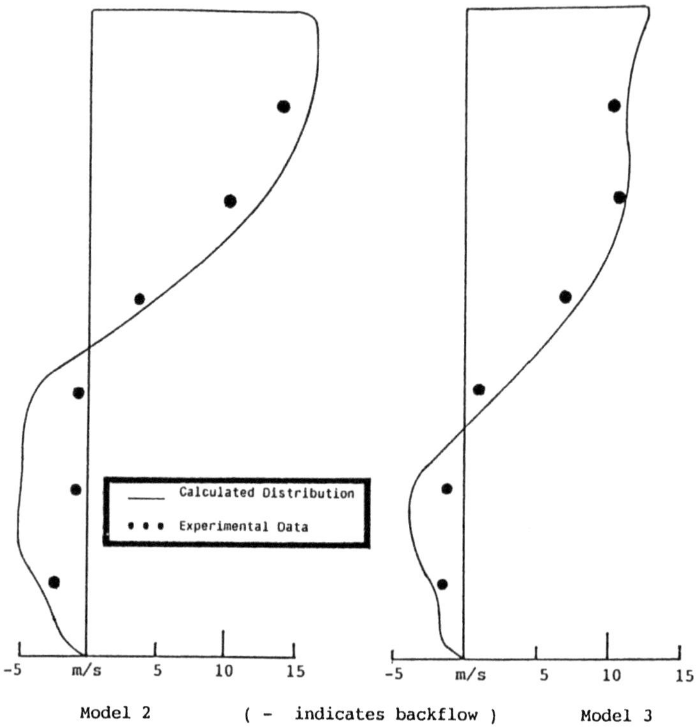

Fig. 4.3.1 Comparision between Calculated and Measured Velocities in the Outlet Center Plane of Model 2 and 3.

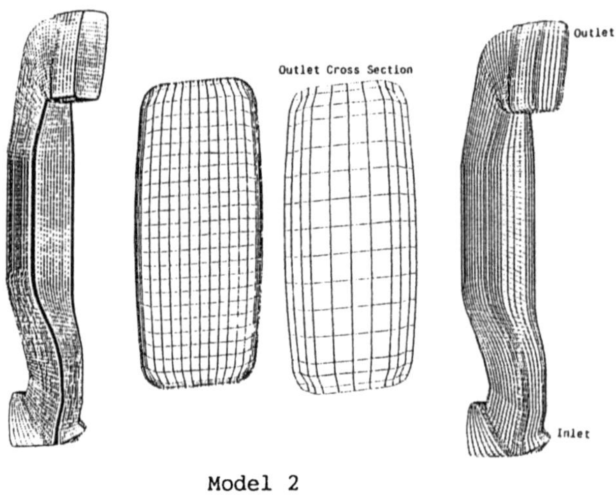

Fig. 4.4.1 Model 2 with 60600 Cells and Model 2c with 15150 Cells and the Outlet Cross Section of both Models.

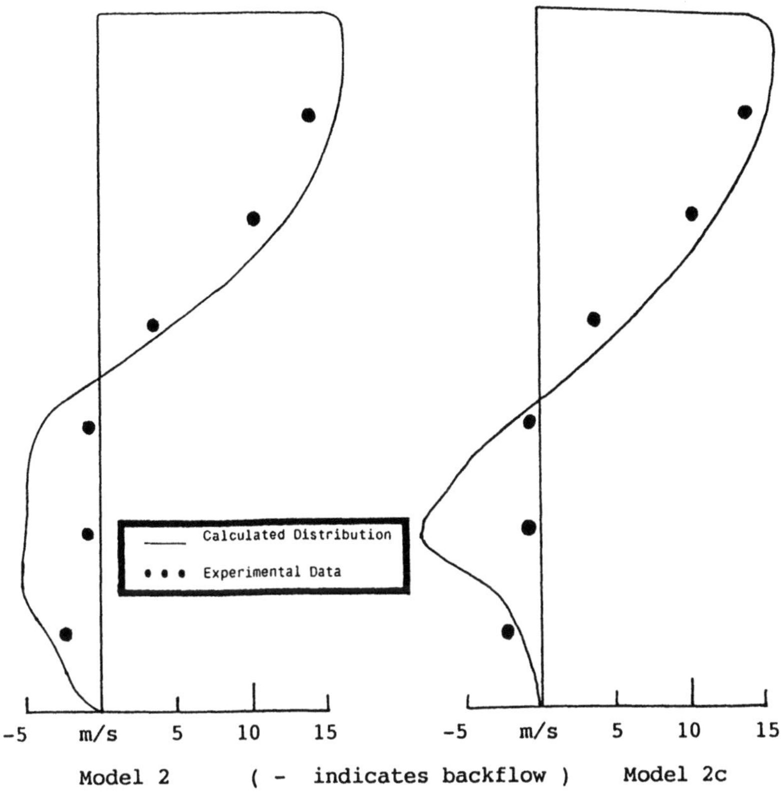

Fig. 4.4.2 Comparision between calculated and measured velocities in the Outlet Center Plane of Model 2 and the coarse Model 2c.

Looking at the comparison of the outlet velocity distributions shown in Fig. 4.4.2, it can be recognized, that the general trend of both results is the same. On the other hand the largest difference occurs as expected for model 2c.

Model No.	Pressure Drop [N/m2]	
	Analysis	Test
2	240	270
2c	220	270

Tab. 4.4.3 Comparision between calculated and measured pressure drop for model 2 and the coarse model 2c.

Tab. 4.4.3 shows the calculated pressure drop of model 2 (60600 cells) and model 2c (15150 cells) compared to the experimental data.
The trend is obvious: A more detailed model improves the results. This is probably due to the ability of the more detailed model 2 to simulate smaller regions of loss than the coarse model 2c can do.

5. Economical Aspect

The development time can be reduced significantly by using CFD-method. In the investigation described here, we achieved a time reduction of 50 % compared with the experimetal method.
Furthermore very detailed information about the behaviour of the flow (pressure distribution, velocities and turbulence quantities) is easily obtained in all locations of the geomertry. These information enables the engineer to optimize his design and to understand the flow.

6. Conclusions

The use of the Computational Fluid Dynamics (CFD) technology to design ventilation ducts turned out to be reliable and successful.

A good correlation with tests was achieved. The difference between test and calculation was always less than 20% considering the pressure drop and the outlet velocity profile. The general trend of all results was always in line with tests.

The analyses can be performed on workstations.

The development time can be reduced significantly by using the CFD-method. In the investigation described here, a time reduction of 50 % was achieved.

The detailed knowledge about the flow behaviour obtained by CFD enables the engineer to optimize his ventilation duct design.
This was practised here and led to a pressure drop reduction of about 60 % and an improved velocity profile at the outlet of the duct.

7. Acknowledgements

All experimental data was supplied from Climate Control Devision (CCD). At this point I specially thank Mr. Ruediger Knab and Mr. Andreas Poessinger, both from CCD, for their excellent support.

8. References

/1/ Benchmarking and Errorestimation of STAR-CD Code, D. Fischer
Ford Internal Report

/2/ Installation and Evaluation of the STAR-CD Code, H. Wolter
Ford Internal Report

/3/ STAR-CD Manual Version 1.813, Computational Dynamics LTD. London

/4/ Numerical Heat Transfer and Fluid Flow, Suhas V. Patankar
Hemisphere Publishing Corporation

/5/ Prediction of the Pressure Drop in Ventilation Ducts using CFD-Technology, D. Fischer
Ford Internal Report

/6/ Approximation konvektiver und diffuser Transportvorgaenge mit Finite-Volumen-Methoden, R. Kessler
NUMET Seminar Erlangen

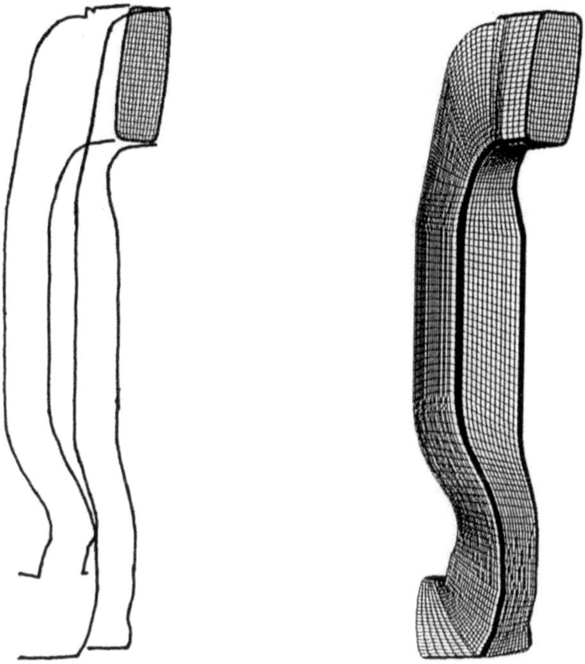

Att. 3.2 Basic CAD Data and Generated Simulation Model

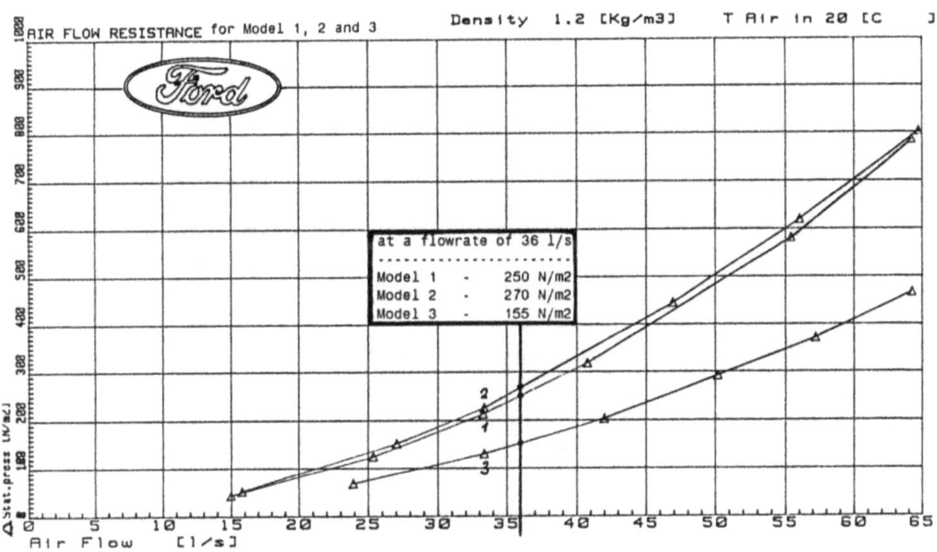

△Stat.press [N/m2] = P (climate chamber) - P (test chamber)
where P (climate chamber) was measured at location Pc shown in Att. 4.1.1
and P (test chamber) was measured at location Pt shown in Att. 4.1.1

Att. 3.4.1 Measured Pressure Drop for Model 1, 2 and 3

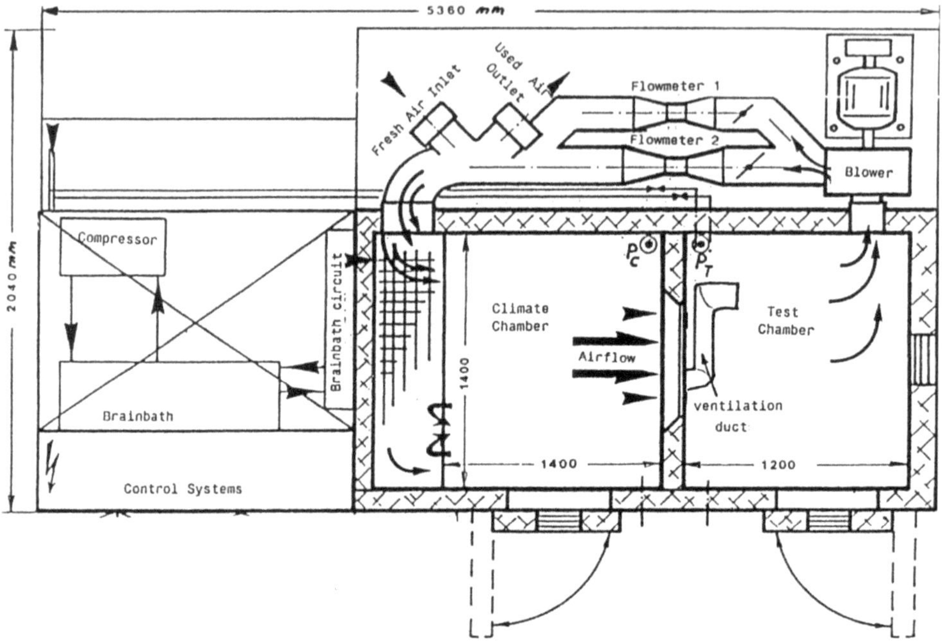

Att. 4.1.1 Test Equipment for Pressure Drop Measurements

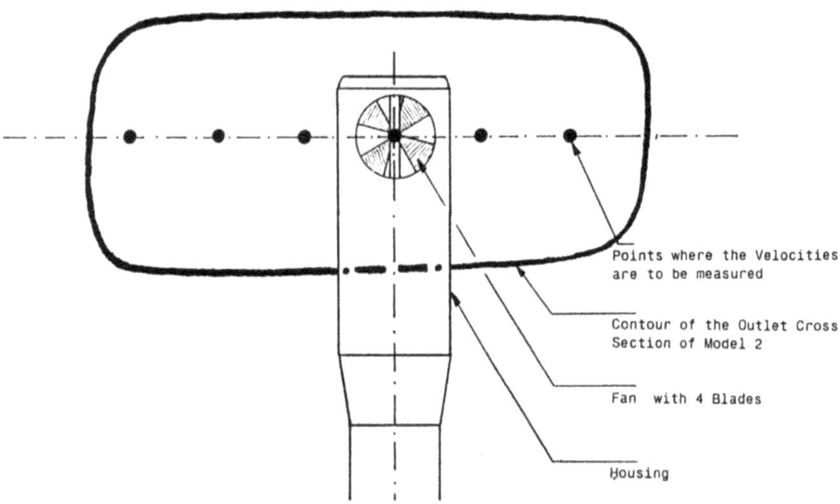

Att4.1.2 Blockage of the Oulet Cross Section due to the Fan Anemometry Velocity Measurements

Software Tools

Interprocedural Analysis: An Advanced Optimization Technique for Large Simulation Packages

Ingbert Graf
Dipl.Ing.(FH), System Analyst, CONVEX Computer GmbH,
Stefan-George-Ring 6, 8000 München 83

Introduction

The supercomputer has become a standard simulation tool for the automotive industry, especially in the area of structure analysis, fluid mechanics and body dynamics. The commercially or customer written simulation packages, generally are written in FORTRAN, and consist of hundreds of subprograms with varying degrees of machine independence. The development, optimization and maintenance of these program systems consumes a lot of resources as time, manpower and of course money. Here an advanced compiler based on interprocedural optimization can be very helpful to improve the efficiency, portability and most of all the reliability of such programs.

For a long time the modularity of high level languages like FORTRAN an C has limited the capabilities of compilers using high level optimization like automatic vectorization and parallelization. Using a conventional compiler, information that is local for one subprogram cannot be used when compiling another subprogram, which may need this local information from the first subprogram. Because of language scoping rules, the compiler must treat each routine separately.

Here we describe the CONVEX application compiler, which goes beyond the normal ways of compilation. This compiler contains a new component called an interprocedural analyzer, that is able to analyze all the relationship of all subprograms of a whole application. This information is used for high level scalar optimization over subprogram boundaries, vectorization, parallelization and for a more intense error checking than conventional compilers and lint-like tools can do. The new optimization techniques are automatic inlining, constant propagation over subprogram boundaries and procedure cloning. In addition to that the "Application Compiler" can do pointer tracking (tracing a pointer over different subprograms and source modules), which is important for optimizing C programs.

Previously most of these optimizations had to be done by hand. The result of these optimizations was a decrease of portability, because changes of the source code had mostly been machine dependent (e.g. calls to specific libraries). The application compiler does these optimization leaving the source code unchanged. Additionally the application compiler behaves like "make", a widely used tool for program development and maintenance, because like "make" it analyzes which module has to be rebuild (recompiled) after changes. This helps the software developer to concentrate on the algorithm rather than on programming techniques.

These new features lead to shorter development and maintenance cycles and more efficiency of simulation software, which is important for developing new products.

General Aspects of Simulation Software Design

Development cycles for cars and trucks are getting shorter and therefore engineering results are needed in a much shorter time. To reach this goal most of the car manufacturers use supercomputers to do simulations which had previously be done in expensive time consuming tests. Numerical simulations are done in most every area of car design and development. For simulating complex engineering problems like, structural and aerodynamic design, engine design and prediction of dynamic behavior of vehicles extensive software packages are needed. The development and maintenance of these packages consumes a lot of manpower and money.
The changing demands of engineering, force software developers to react in a short time, creating new software packages or adapt old packages to new problems. The rapid changes in computer hardware lead to a continuous porting and optimization process so that new hardware can be used efficiently. Tools available in such an environment help the programmer to write more reliable code and enable the software developer to concentrate on the solution of the engineering problems and the development of efficient algorithms, rather than on computer related problems like debugging, porting and optimizing.

One aspect of software design to write better code is modularity. A clear program structure, which can be reached by splitting code into logically independent modules, increases the readability of the program code and eases the adoption of new or changed parts. It also makes it possible to easily add user defined new modules to a complex program.
Well defined interfaces between the modules open the possibility that several developers can work on the same program without big interference effects.

On the other hand this modularity makes an automatic optimization much more difficult, because of the limited range of source code a compiler can look at.

But the need for an increased number of more complex simulations on high performance computers requires highly optimized software, that uses all of the features of that modern computer hardware. This requires compilers which can automatically produce highly optimized code and overcome old hand optimization. New compiler technologies can reach a degree of optimization which cannot be reached by hand.

Another aspect of software design is the reliability of the results produced. FORTRAN and C, which are the high level languages mostly used give the programmer much freedom in writing interfaces between different modules. The type of arguments or global values in different subprograms can be of different type and number. Conventional compilers lack of the ability to check if interfaces have inconsistencies. But exactly these inconsistencies cause a lot of problems in most every software package.
Specially old packages which have been ported from FORTRAN IV to FORTRAN 77 can contain a lot of potential for such errors.
A point which is becoming more and more important is the portability of programs. Heterogeneous hardware environments (PC's up to supercomputers) found in most automotive companies require a high degree of this portability. This means, for example, a software package that is developed on a workstation must run on a supercomputer. Basis of good portability is the commitment to standards, no dirty programming. But what to do with old programs? In the past a lot of programming tricks have to be used because of hardware limitations of older computer systems. Exactly this way of programming causes a lot of problems when such a program has to be ported to another computer system.
Costs for debugging and porting increase with the size of a program and can make up more than half of the amount that has to be spent for developing a software package. Using high performance development tools can reduce these costs dramatically.
To meet all the described requirements of software development a good Software-Development-Environment is needed. Tools tailored for modern needs of software development have replaced old batch like development procedures. The UNIX operating system offers a lot of these tools like make, awk and lint, to name only a few.
Central part of Software-Development-Environment is a high performance compiler. The quality of such a compiler has a strong influence of the quality of the software package developed with this compiler. Such a compiler can no longer be looked at, as a tool for converting High-Level-Language (HLL) to machinecode. Within the last years the tasks of compilers have grown.

Hardware is getting more complex (pipeline architecture like Vectorprocessors or new RISC processors; parallel processors) and the siz of program that have to be compiled reaches some hundreds of thousands of lines.
Programmers now expect that the compiler produces the most efficient machinecode possible for those complex hardware architectures, plus a high level error checking, all integrated into one easy to use development environment.
Figure 1 shows the chronology of high optimizing FORTRAN compilers.

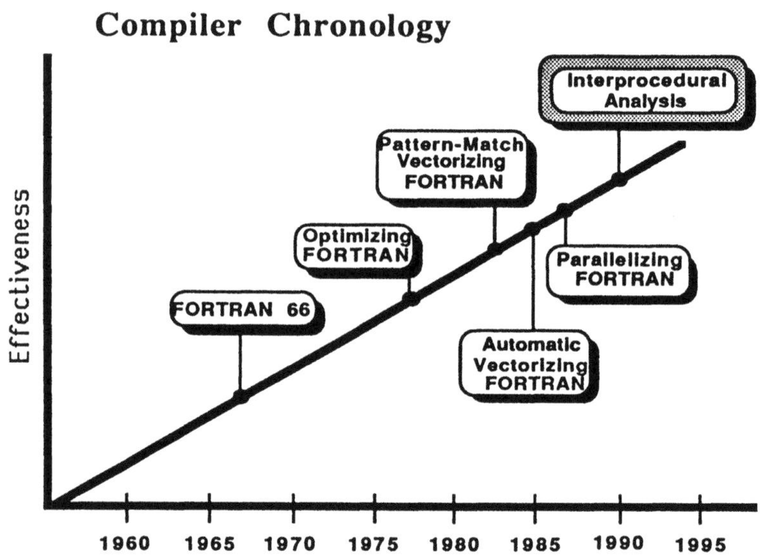

Fig. 1: Chronology of FORTRAN Compiler development

Interprocedural analysis and optimization

Most of the compilers use a dataflow representation of the HLL routines to apply optimization techniques like automatic vectorization, automatic parallelization, constant folding, strength reduction, scalar propagation, common subexpression elimination, invariant code motion , dead code elimination and induction value removal. But with conventional compilers the scope of the compilation spans only over one routine. Information from other routines are not used.
Interprocedural analysis examines all routines of an application. A dataflow representation of each routine is built and stored into a database. From this information a dataflow structure containing all information from the whole application is built. State-of-the-art optimizations including all optimization

found in conventional compilers and interprocedural optimizations are applied. New features of interprocedural optimization are automatic inlining, constant propagation over subroutine boundaries and procedure cloning. The fact that the optimizer can use more information also opens the possibility to apply more and better conventional optimizations techniques. The interprocedural optimizer examines memory locations and relationships between variables, arrays and functions and preforms pointer tracking to find complex relationships of different memory locations. This information can be used for a better detection of data dependencies. Data dependency means that data created in one pass of a loop iteration is referenced in one of the subsequent iterations of the same loop. This data dependencies often prevent vectorization and parallelization. In most cases the interprocedural optimizer can use loop transformations to get around such dependencies, if enough information about the loop parameters is available. In some cases conventional compilers detect data dependencies which only look like a dependency.
In the following example a conventional compiler cannot automatically decide if a dependency exist or not:

conventional compiler :

```
PROGRAM MAIN
PARAMETER (N=100,M=5)
.
CALL NO_DEP(A,N,M)
.
.
END


SUBROUTINE NO_DEP(X,N,M)
REAL X(*)
.
DO I = 1,N
    X(I) = X(I+M)*A(I)
ENDDO
.
RETURN
END
```

interprocedural compiler :

```
PROGRAM MAIN
PARAMETER (N=100,M=5)
.
CALL NO_DEP(A,N,M)
.
.
END


SUBROUTINE NO_DEP(X,100,5)
REAL X(*)

DO I = 1,100
    X(I) = X(I+5)*A(I)
ENDDO

RETURN
END
```

Using information from the calling routine would make it very easy for the compiler to decide. With conventional compilers the programmer must instruct the compiler with directives if an apparent dependency should be ignored.

Automatic Inlining
Inlining is an optimization technique which replaces the call of a subprogram with the text of the subprogram itself. The call overhead for frequently used subprograms can be eliminated by using this technique. Modern compiler technology provides algorithms to determine calls to subprograms which are profitable to inline. A general rule for inlining is the size of the subroutine should be small and the number of calls to this routine should be high.

Constant Propagation
Constant is a well known Optimization technique which means that the symbolic name of a constant is replaced by the value of that constant expression or that a constant expression is evaluated at compile time and the constant value is used by the compiler instead of the expression. This gives the advantage that computation can be already done at compile time and also a better vectorization and parallelization can be achieved.
looking at interprocedural optimization constant propagation is extended to the whole application. Such constant values can appear in argument list and global values. Propagating these constant values can improve vectorization and parallelization by improved resolution of recurrences and a greater opportunity for loop transformations, can also improve automatic inlining and can help to eliminate dead code which is not be found by conventional compilers. The interprocedural algorithm propagates entities like literal constants passed as arguments, variables passed as arguments that always have the same value on entry to a procedure, global variables assigned in procedures which are called before a given procedure and global variables which are statically initialized.

Examples for resolved recurrence using constant propagation:

```
PROGRAM MAIN                  SUBROUTINE SUB1(X,N,M)
PARAMETER (N=100,M=5)         REAL X(*)
REAL A(1000)                  DO I = 1,N
CALL SUB1(A,N,M)                 X(I) = X(I+M)
.                             ENDDO
.                             RETURN
END                           END
```

```
PROGRAM MAIN                  SUBROUTINE SUB1(X,Y,N,M)
PARAMETER (N=100,M=5)         REAL X(*),Y(*)
REAL A(1000),B(1000)             DO I = 1,N,M
CALL SUB1(A,B,N,M)                   X(I) = X(I+1) +Y(I)
.                             ENDDO
.                             RETURN
END                           END
```

Procedure Cloning

Procedure cloning, which is a new feature of interprocedural optimization, can be a very effective optimization technique. Using this technique the compiler can make several copies(clones) of a subprogram, which can then be optimized differently. This can be used when a subprogram is called at different places with different parameters that wold require a different kind of optimization. For example, vectoroperations on very short vectors (1-3 elements) may not be vectorized while longer vectorlengths shold be vectorized. If the vector length can be determined at compile time the compiler can make 2 copys of that subprogram, vectorize one and compile the other for scalar execution.

```
PROGRAM MAIN
PARAMETER (N=100)
REAL A(100),B(100)
CALL SUB1(A,B,N, 1)
CALL SUB1(A,B,N,-1)

END
```

```
SUBROUTINE SUB1(X,Y,N,M)
   REAL X(*),Y(*)
   DO I = 1,N,M
      X(I) = X(I+1) +Y(I)
   ENDDO
RETURN
END
```

The compiler creates a vectorized version of sub1 and replaces the second call of sub1 with a call to a clone of sub1 which will not be vectorized because of the recurrence resulting from the negative step in the DO loop.

Interprocedural Error Checking

The stored information can also be used for extensive error checking which includes all the semantical and syntactical checks conventional compilers use. Using conventional compilers, language scoping rules prevent the compiler to find programming errors in the interface between routines or errors that span over several subprograms. With the extended information of interprocedural analysis, the compiler can detect more errors, which leads to a more reliable and stable application and saves hours of debugging. This also shortens the time to validate a software product.

Errors that belonging to this category are mismatched number and types of formal and actual arguments, mismatched types of function calls and return values, scalar actual arguments passed to array formal arguments, Constants passed by reference to formal arguments which are assigned, FORTRAN dummy variables that are aliased to COMMON variables, FORTRAN dummy variables that are aliased to each other, array access beyond array boundaries, uninitialized global variables or arrays, defined-but-not-called procedures and hidden aliases from global values to arguments passed to subprograms.

A wide spread source of runtime errors is the inconsistent use of COMMON blocks in FORTRAN. Conventional compilers cannot detect these errors, because it examines only one module at a time. Interprocedural analysis finds such inconsistencies. Equivalent problems can be found in parameterlists in FORTRAN and C.
In the past a lot of manpower was lost in the detection of such programming errors.

Example for hidden aliasing :

```
PROGRAM MAIN                 SUBROUTINE SUB1(X)
COMMON A                     COMMON A
CALL SUB1(A)                 A=1
.                            X=2
.                            RETURN
END                          END
```

Another type of programming errors are variables which are used but not initialized. On most computers these variables a value of zero is assigned. This can lead to runtime errors and in the worst case to wrong results which in some case where never detected.
Interprocedural analysis gives also information about subprograms which are never used. Such routines often exist in big software packages containing some hundreds of subprograms .

Pointer Tracking
Pointer tracking is also a new feature of interprocedural analysis. It determines if pointers are aliases to the same memory location. The compiler tracks every memory location a pointer points to. With this information the compiler can then decide if loops using pointers are save to vectorize, which means that there are no dependencies between vectorelements. A conventional compiler must assume that global values and formal arguments which are declared as a pointer can point anywhere in memory which means that there is always a potential data dependency between two pointers. This fact prevents vectorization and parallelization of loops containing references to such pointers. The interprocedural optimizer uses all the information to remove potential recurrences that are not true recurrences, and so enables vectorization and parallelization of loops. This leads to a dramatic increase of vectorizable loops in C that cannot be automatically vectorized by normal compilers. In FORTRAN pointer tracking can be used to detect hidden aliases described above.

Example of the vectorization of pointers:

```
char *malloc();
double static1[200];

main()  {
    double automatic1[200];
    double *p, *q, *r;
    int i;
    p = static1;
    q = automatic1;
    r = (double *)malloc(200 * sizeof(double));
    for( i=0; i<200; ++i )
            *p++ = *q++ - *r++;
}
```

CONVEX Application Compiler

Interprocedural optimization research for this compiler was done at Rice University and CONVEX. The CONVEX Application Compiler is the first production compiler using such kind of new technology.
This new compiler incorporates two new parts. An interprocedural analyzer and optimizer together with a make like development environment. Figure 2 shows the data flow during the compilation process. The Application Compiler is based on existing CONVEX compilers.

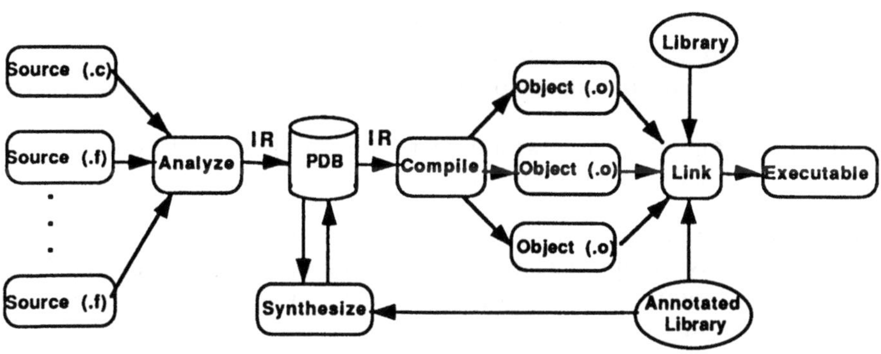

Fig. 2 : Dataflow during compilation

Although the CONVEX Application Compiler uses all possible optimization techniques, it will not perform optimizations that other compilers should not perform. It works with software tools already available on CONVEX systems, such as editors (vi, edt, emacs), source control systems (RCS, SCCS) , profilers (prof, gprof, CXpa), and debuggers (csd, CXdb).
The compilation process is controlled by a program called build. Build behaves and looks a lot like make. The rules how to build an application are stored in a buildfile. Although the concept of build is analogous to make, the use of build is much simpler. The structure and syntax of a buildfile is easy to understand. The order how files are compiled is determined by build itself while it scans all the files belonging to an application and has not to be specified by the user.

Examples of a buildfile :

```
.   -O2                       ! compile all sources in the current directory with -O2
main.f   -O1                  ! compile main in the current with -O2
link FORTRAN -lveclib         ! link all modules plus routines from vectorlibrary
```

```
FFLAGS = -pd8                 ! Macro known from make
foo.f -O2
bar.f -O2
main.f -O1 -or none
link FORTRAN timer.o
```

Build knows most all of the macros that can be used in a makefile. An automatic conversion from existing makefiles to a buildfile is possible using a utility belonging to the application compiler. The CONVEX application compiler can be used in place of make , or it can be called from make.

To control the compilation process a numerous set of commands, options and directives can be used to override the default actions of the compiler.

The interface structure of all system libraries routines is known by the compiler. It can savely optimize calls to these routines and detect interface errors. If such an interface problem is found the compiler issues warning or error message like this :

```
ERROR!: Argument 2 of SUB1 has inconsistent type in test.f on
line 10 and sub1 on line 1
Call to SUB1 on line 10 of test.f : (INTEGER*4, REAL*4)
Defn of SUB1 on line 1 of sub1.f : (INTEGER*2, REAL*4)
Fatal type conflicts encountered.
```

If external object modules are used extra information can be provided. These informations are called annotations and contain interface definitions and side effects information which can then be used by the application compiler to apply correct optimization at the highest degree. These annotations contain only source code to describe the procedure header, argument declarations and informations about the side effects of a subprogram.

Example for an annotation in FORTRAN :

```
SUBROUTINE SAXPY(N, A, X, INCX, Y, INCY)
REAL*4 A
REAL*4 X(*), Y(*)
INTEGER*4 N, INCX, INCY
C$DIR PSUM_NO_GLOBAL_MODS
C$DIR PSUM_ASGS(Y)
END
```

and C :

```
function header
    {
    side effect descriptions
    return;
    }
```

These annotations are transformed to internal representation by a utility and stored in the database for use by the interprocedural analyzer and optimizer.
The use of the application compiler also eases the combination between FORTRAN and C. By default FORTRAN passes arguments by reference (passing a pointer to an argument), while C passes arguments by value. This fact can cause interface problems because none of the conventional compilers detects if there is a type mismatch between arguments.
An additional feature of the application compiler the printout of a call graph of an application. This information can help a lot to understand the structure of an application.

Here are some parts of a logfile produced by the CONVEX application compiler :

```
Call Graph:
BODY
    SQ2
    SGECO
        SASUM
        SGEFA
```

```
            ISAMAX
            SSCAL
            SAXPY
        SSCAL
        SDOT
        SAXPY
    SGESL
        SAXPY
        SDOT
    SUB2
        SQ2F
    SUB3
    SUB4
    SUB5
        BLCONT
            SPLINE
            LAMINR
                FFK
            TURB2
```

Detection of dead code :

```
Analyzing procedure 'BLCONT'
Optimization: Code from line(s) 29-38 of blcont.f unreachable
```

Constant propagation:

```
Performing constant propagation
Constants on entry to procedure 'SUB5': JREAD=5, JWRITE=6,
KFILE1=7, KFILE2=8, KFILE3=9, KFILE4=10, KFILE5=11
```

Automatic inlining:

```
Optimization: Marked call to ISAMAX on line 402 of lin.f for
inlining
```

Called but not used subprogram :

```
Warning: Procedure 'VGX' is defined but not called
```

Detection of a hidden alias:

```
ERROR!: Assignment to 'N' in 'SUB1' is invalid due to alias
ERROR!: Assignment to 'ETA' in 'SUB1' is invalid due to alias
with GAMMA
```

If the interprocedural analyzer can evaluate which subscripts of an array are referenced it prints them into an array summary for each routine.

Here is an example of an array summary for a routine :

Routine PFP3

VARIABLE	USE	ASG
COEF	(*)	
S	(*,1:NP)	(*,*)
S1K/SN/	(*)	(*)
VIP	(1:NP)	(1:NP)
_blnk/XNP/	(1:NP)	

Summary
The CONVEX Application Compiler is the first compiler product using interprocedural optimization techniques. It offers the highest degree of state-of-the-art optimizations together with a broad range of error detection mechanisms. It can be used on every CONVEX system (C2 and C3 series) including the CONVEX PREEMPTOR 5000 realtime systems.

References :

Smith P.: Introduction to the CONVEX Application Compiler
CONVEX internal paper

Metzger R.: The CONVEX Application Compiler
CONVEX internal paper

CONVEX Application Compiler User´s Guide, Document No. 720-004030-000
(September 1990),CONVEX Comp. Corp., Richardson,Tx.

CONVEX Application Compiler Release Notice, Document No. 720-004130-001
(April 1991), CONVEX Comp. Corp., Richardson,Tx.

Simulation Workbench – An Integrated System for Complex Multiprocessor Simulation Environments

by Albert Sölter, Concurrent Computer GmbH, Planegg - Martinsried.

I would like to thank my colleagues in the USA, especially Charles Fung and Jem Treadwell for their contributions to this paper, and as well all the people for their patience in answering my questions that made it possible to prepare this paper even before any press or other public announcement of this new Software product was made.

Abstract:

System integration of real-time frequency-based simulation applications has long been a difficult and time-intensive task. A simulation application comprising several modules is often developed by different groups of programmers. Each simulation module needs to be executed at an application specific frequency and must complete its execution before it is scheduled again.

When modules overrun their time constraints, the simulation needs to be adjusted. For the systems integrator, this can often be a difficult process. Not only does the application need to be modified, it must be recompiled and relinked as well. Once this is done, the systems integrator must trace when each module executes, and whether the new scheduling scheme efficiently uses the dedicated processors. As long as there is a need to test varying configurations, the systems integrator will have to continually repeat this manual process.

Concurrent facilitates the task of system integration through the use of a tool called the SIMulation Workbench [1] (Sim WB). Through its window-based user interface, the SIMulation Workbench allows the integrator to configure, start, suspend, reconfigure, resume or stop the simulation without the need to write a single line of code. A back end code generator automatically creates the required code that schedules the execution of the application's modules (written in C, FORTRAN or Ada). Any new modules that need to be added cause the SIMulation Workbench™ to automatically generate new code and relink it with the other modules. By providing this kind of support, the SIMulation Workbench allows the systems integrator to concentrate on the simulation design without being burdened by low level coding.

In addition to integration support, the SIMulation Workbench also collects simulation timing information and, for each module, reports the percentage of processor time used in the application. If a module cannot finish its task within the time allocated, the SIMulation Workbench allows the systems integrator to dynamically change, reconfigure and reassign modules to allow for a more balanced system. The automation of such system integration activities can result in a much shorter development cycle for most real-time frequency-based application.

---

1 SIMulation Workbench is a trademark of Concurrent Computer Corporation

1. The Problem, why simulation is no fun anymore

During the life cycle of any given simulator, the model used, because of changes to the vehicle being simulated, increases significantly in complexity. Historically, at some point in time, the total view of all tasks, modules, software parts and timings throughout the total simulation environment is lost. Someone, whether the operator, the system manager, or the system integrator has to control all of this information and adapt them to all types of outside environments and tests, and to implement additional tasks, change timings and assemble that particular simulation system as needed and used in the next run.

Then the next step is, during the run, assure that all timing limits are enforced and controlled, that any particular part of the whole model meets the requirements, that any scheduling is done at the right point of time. Now, on a step-by-step basis the modules and tasks have to be modified, moved between CPUs, have to be adapted to the time requirements of the simulation model, have to be tested, tested and tested again. Then ....

But all and everybody, using, developing, integrating such type of simulators knows about this job and the difficulties that occur with each modification, each extension.

So: **It is time for a better solution............**

2. Introduction, light at the end of the tunnel

Generally, a typical frequency-based simulation application consists of a number of program modules that are scheduled at specific frequencies required by the application. Real-Time constraints require each module to finish its execution before the next scheduling cycle starts. Modules that overrun their allocated time frames need to be terminated and such occurrences have to be reported as errors.

Such applications normally have to be tuned by adjusting the frequencies of invocation of the modules until a balanced system is arrived at.

Concurrent's Real-Time UNIX[2] (RTU[3]) provides kernel support, at the OS process level, for frequency based scheduling through Synthetic Period Scheduling (SPS). SPS allows a process (or thread) in RTU to specify the dynamics of its desired execution pattern which will be synchronous to a given frequency.

The Simulation Workbench[4] (Sim WB for short hereafter) is developed to provide user level software support tools for automating the design and implementation of simulation applications that utilize the SPS facility. Sim WB automates the scheduling of modules at both the OS process and subprogram unit levels.

---

2 UNIX is a registered trademark of ULS (UNIX System Laboratories, Inc.)

3 RTU is a trademark of Concurrent Computer Corporation

4 SIMulation Workbench is a trademark of Concurrent Computer Corporation

3. The underlying Hardware, something you can count on

Concurrent Computer manufactures standards - based computers for Real-Time applications. All types of machines are multiprocessor systems,working on a global shared memory over a privet system bus to solve the needs for high system throughput. One or two separated I/O-buses (VMEbus[5] or VME-64-bus) make these system architecture the design for simulation applications. Special on-board integrated clock devices generate interrupts and complete the underlying Hardware.

4. Real-Time-UNIX, an OS that makes the hardware work

UNIX as a standards - based operation system is becoming popular in more and more markets Where as a standard UNIX and the requirements for Real-Time are actual opposite requirements, RTU (Real-Time-UNIX) is designed to solve the needs of the Real-Time-community. True symmetric multiprocessing on tasks and threads, process- and kernel-preemption, real-time-scheduling, frame-based scheduling, memory locking of processes, modules, threads and data, dedicated device allocation to all types of codes, an extended mechanism for Hardware and Software interrupts as ASTs[6] (Asynchronous System Traps) are only some of the Real-Time extensions built into the OS-code.

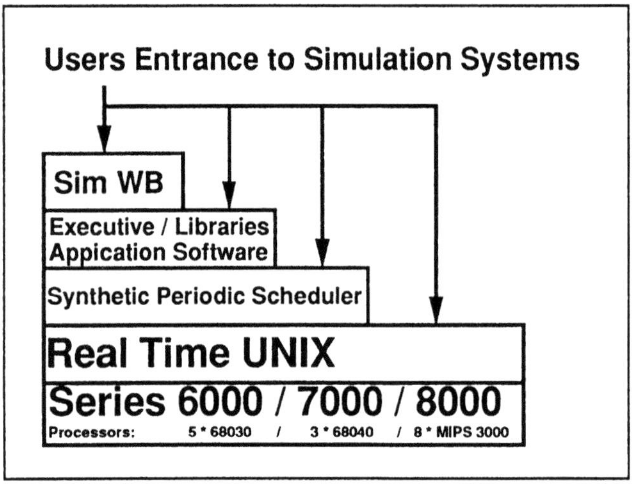

Concurrent provides Multi-Level Access for Simulation System Builders

5. Synthetic Periodic Scheduler, basics for a frame-based simulation

In true simulation environments running tasks in the computer are not scheduled continuously, but require a repetitive pattern of execution over time, to synchronize

---

5 VMEbus is a trademark of Motorola Corporation

6 Remark: There is a list of terms and definitions used at the end of this paper

with other tasks of the outside world. The Synthetic Periodic Scheduler (SPS) in a scheduling mechanism providing this kind of functionality to the RTU operating system. The SPS is expressly tailored to support applications requiring synchronized pattern-based scheduling, also referred to as frame - based or frequency - based scheduling.

The Synthetic Period Scheduler provides a task with support for:

- Periodic scheduling of a task according to an elected execution pattern, including phase synchronization across tasks, processors and machines
- Detection and notification of execution overrun, including abort and shutdown handling for degenerate conditions
- Access to runtime statistics, performance, and dynamic timing data

Using the SPS, an execution pattern for a task is specified as a function of a given frequency. The period of the frequency is called a minor cycle. A minor cycle is the basic period of time during which a participating task may be scheduled. Under the SPS scheduling mechanism , the base scheduling frequency and minor cycle, a task may be specified as active or dormant.

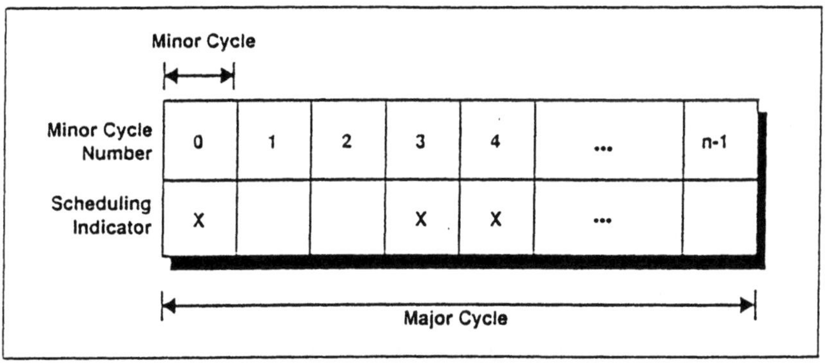

Task Execution Model

Minor cycles are grouped into a fixed repeating sequence called a major cycle. A major cycle consists of the pattern of minor cycles repeated over time. The execution of a task during a major cycle corresponds to the pattern of minor cycles during which the task is active or dormant.

A task will be scheduled to run at the beginning of an active minor cycle, and will commence execution when it is the highest priority runable task, perform its functions, and then relinquish control of the processor before the expiration of the current minor cycle. Failure to relinquish control before the beginning of the beginning of the next minor cycle may, if desired, result in raising an overrun condition.

A private scheduling map is created for each CPU participating. The scheduling map is per-CPU to minimize memory contention latencies that would be present with only one global map. The scheduling map contains a fixed number of scheduling slots, each of which represents the major cycle of a registered task. Using the major and minor cycles specified in the scheduling maps, the SPS synthesizes the execution pattern for the registered tasks.

Scheduling occurs as the result of an event visible to the application task, rather than through preempt / resume mechanisms. Several processors may be participating in the SPS, and a mechanism is provided to propagate the scheduling event such that the SPS is operating virtually simultaneously on all processors.

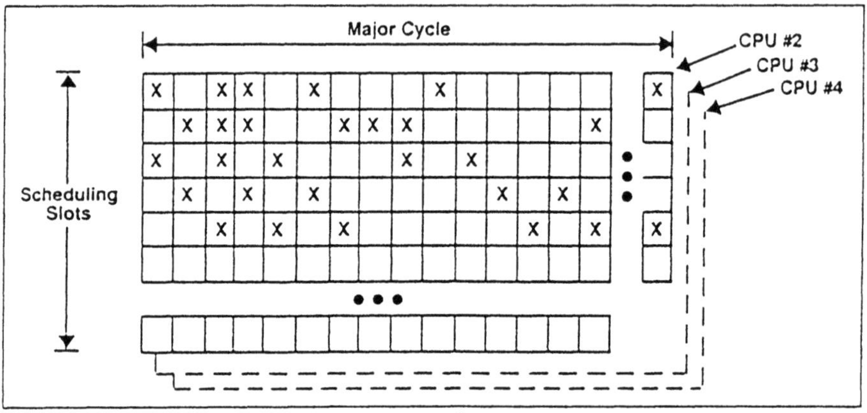

SPS Runtime Model

The SPS might be connected to an outside arbitrary interrupt source and so to custom timing hardware. If internally connected SPS is sourced by the DUART clock, available on all Concurrent CPUs, supporting a frequency range from 30 minutes to 1ms.

6. **About SIMulation Workbench**, you see what you get

6.1. **The Features**, bits and pieces

The Simulation Workbench will provide users with an automated environment for designing and implementing frequency based applications that utilize RTU's SPS. This will allow the the application designers to concentrate their efforts in designing the individual application subsystems. Sim WB is designed to relieve the application designer of the tedious house keeping chores such as:

- writing code to interface with SPS,
- writing code to implement subprogram scheduling,
- writing code to detect overruns and take appropriate action(s) , and
- modifying code to alter the parameters of the simulation in order to arrive at a well balanced system.

It has a graphical user interface (GUI) built on X Windows[7] and Motif[8] . Operation of this interface is mostly menu - driven and mouse - oriented. The figure shows a high level overview of the architecture of the Simulation Workbench system.

---

[7] X Window System is a trademark of MIT

[8] Motif is a registered trademark of the Open System Foundation, Inc.

Functionally, Sim WB can be broken down into four major portions:

- Application Configuration Manager
- Executive Generator
- Control Monitor
- Application Launcher

These functional modules are discussed in more detail later on.

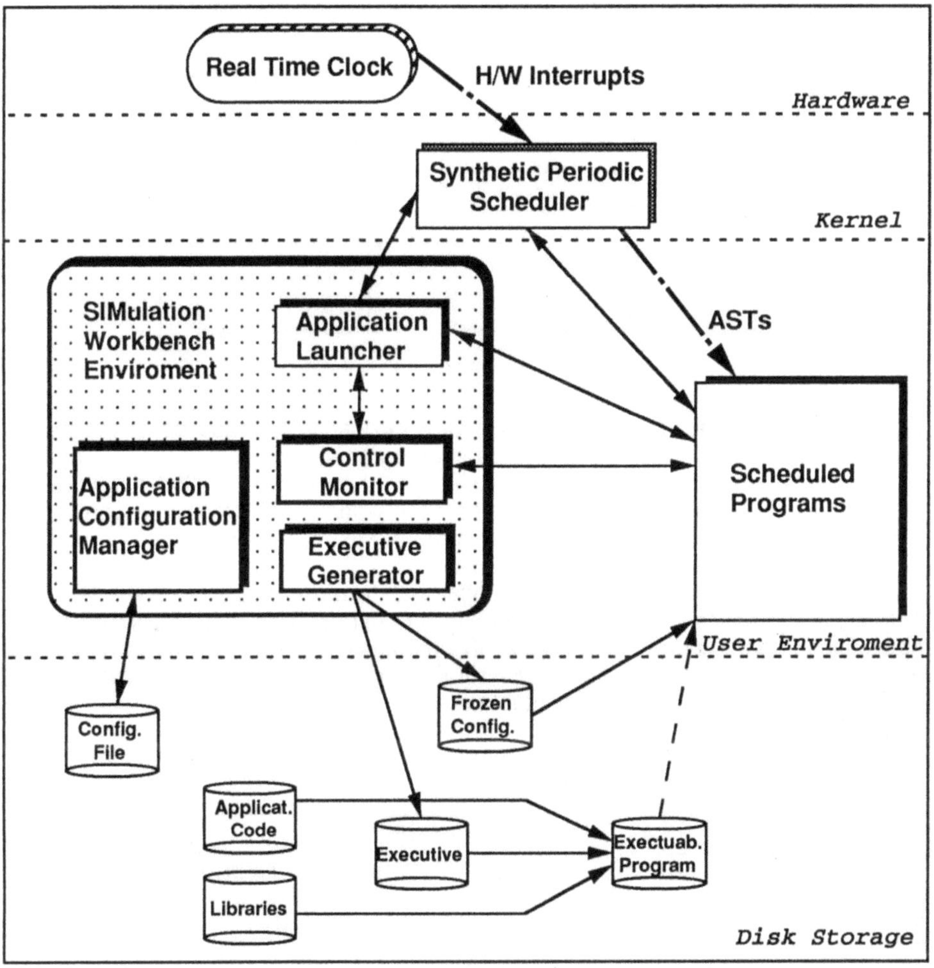

Simulation Workbench- Theory of Operation

6.2. Theory of Operation,the way the parts work together

Sim WB takes a high level description of the characteristics of an application and automatically generates the executive code to schedule user-provided program modules at specified frequencies.

There are two possible views of a Sim WB application-static and dynamic.

The static view of an application is its programmatic structure that ties the user program modules together with the executive modules generated by Sim WB to form the executable files at compilation time.

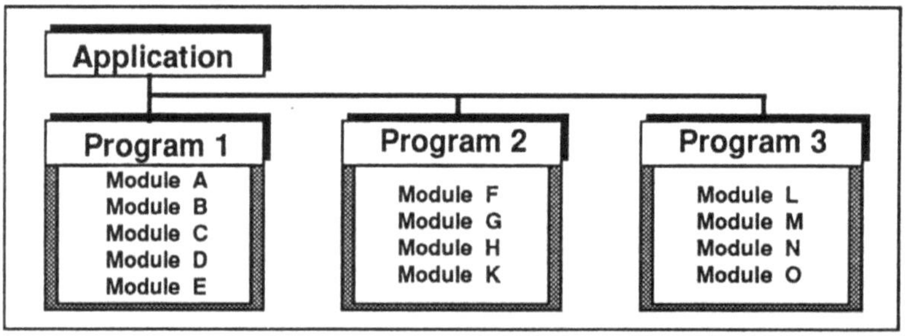

Static View of Simulation Application

As shown in the above diagram, a Sim WB application is made up of programs, each having a set of modules (e.g., Ada subprograms) defined within the program. For example, Program 1 might be a collection of subprograms that simulate the body behavior, while Program 2 implements the wheels and Program 3 consists of modules that control the motor characteristics.

The above static structure of the application has to be registered with the Sim WB before the dynamics of the application can be defined. This static structure of the application is hard coded into the executive modules generated by the Sim WB and therefore cannot be easily altered without the regeneration of new executives and recompilation of the program executables.

The next diagram shows a possible dynamic view of a Sim WB application based on the static structure depicted above.

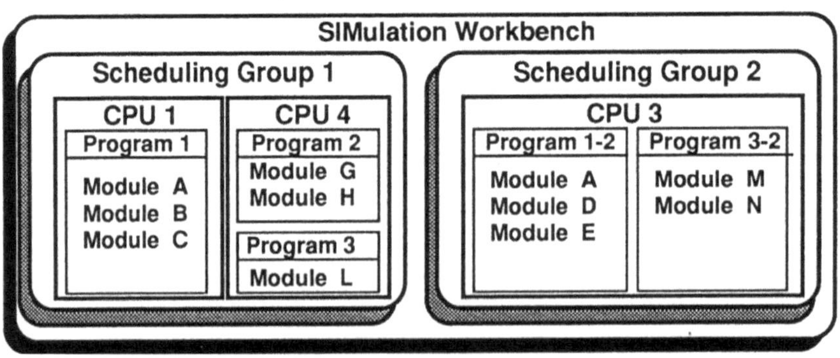

Dynamic View of Simulation Application

Under the control of Sim WB, two schedulers are created for the two scheduling groups. Scheduling Group 1 drives the scheduled programs executing on two dedicated processors, namely CPU2 and CPU4, at a base frequency of, e.g., 60 Hz. Scheduling Group 2 drives the scheduled programs executing on one dedicated processor CPU3 and might be based on a base frequency of 50 Hz.

Program 1, an instance of execution (scheduled program) derived from the statically defined program Program 1, is scheduled on CPU2 by the scheduler Scheduling Group 1. This particular scheduled program only calls modules Module A, Module B and Module C out of the entire set of "registered" modules in Program 1.

The executive module of Program 1 schedules subprogram calls to Module A, Module B and Module C at appropriate minor cycles when scheduled program Program 1 becomes active on CPU2.

On the other hand CPU4 has two scheduled programs running on it, namely Program 2 and Program 3. Scheduled program Program 3 might be specified to be a scheduled "background" program that runs only if there is idle processor time on CPU4. Likewise, CPU3 also has two scheduled programs running on it. These scheduled programs are driven on CPU3 by the scheduler of the Scheduling Group 2 because their subprograms needs to be invoked in synchronization with a 50 Hz clock.

It is worth noticing that scheduled programs Program 1 and Program 1-2 are different instances of execution of the statically defined program Program 1, and they run on different processors under the control of different scheduling groups. They are distinguished from one another by their different scheduled program names.

This dynamic structure of the run-time characteristics of the Sim WB application can be easily altered on the fly, and no recompilation and relinking of the programs is required as long as the static structures of the application programs do not change. For example, during a trial run of the application with the above dynamic structure, we might find that CPU4 is logging an average utilization of close to 100% and the background program Program 3 is probably left with very little processor time to do its jobs. To balance out the load, we might want to move Program 3 from its current position on CPU4 over to CPU2 and schedule it after Program 1, provided that Program 3 does not have a dependency on the execution of Program 2. later we might add module Module F to be scheduled in Program 1-2 when a new idea needs to be tried out in our application.

Alternatively, we might want to juggle the execution sequence (priority) of modules within Program 1-2 from its current Module A - Module B - Module C to Module A - Module E - Module D because we made a mistake in the first cut.

It is in these kind of restructuring of the dynamics of a frequency-based application that the Sim WB wins an upper hand. The user can instruct Sim WB to change the dynamics of the application through the use of a friendly graphical user interface and such changes are effected by Sim WB automatically on the fly - even after the application has already been started.

The following sections describe the functionalities of the various parts of the Sim WB in a little more detail.

6.3.1 Application Configuration Manager, the system needs to know...

The Application Configuration Manager portion of Sim WB serves to establish a configuration of a given Sim WB application.

The major responsibilities of the Application Configuration Manager includes:

- Loading a configuration file from disk and displaying the configuration information graphically in a hierarchical, easy-to-understand manner.

- Allowing the application designer to navigate his way through the hierarchy that describes the configuration of the application being developed, making modifications to the configuration as he sees fit.

- Saving the configuration back into a disk file for later use.

6.3.2. Executive Generator, control code for the application

The executive generator automatically generates the executives which consist of top level control code responsible for all the housekeeping work necessary to effect the periodic execution of scheduled modules, as well as the collection of application performance data. The application designer is responsible for compiling and linking these executive modules with his own application modules to produce the program executables.

The executives are generated in Concurrent´s C3 Ada. Any combination of Ada, C or FORTRAN simulation modules can be driven by these executives. Proper procedure or function call interface will be automatically built into the executive to make calls to user-provided Ada, C, FORTRAN, or assembly modules.

6.3.3. Control Monitor, you get what you see

As the name implies, the Control Monitor of Sim WB serves two major functions, it controls the execution of the application and it also monitors its execution to gather and display timing statistics. The following sections describes these functions in more detail.

6.3.4. Application Launcher, control for the software booster

The Application Launcher is a separate process started by the SPS Manager that does most of the actual work of managing the real-time application. This tool is normally transparent to the user except when the need arises to run the simulation developed by the Sim WB in a standalone mode, without the graphical interface of the Sim WB.

6.4. The User Interface, you see, what you need to use

Today one of the most critical parts to make such a computer system valuable to the user is the user interface, in this case the graphical user interface (GUI).

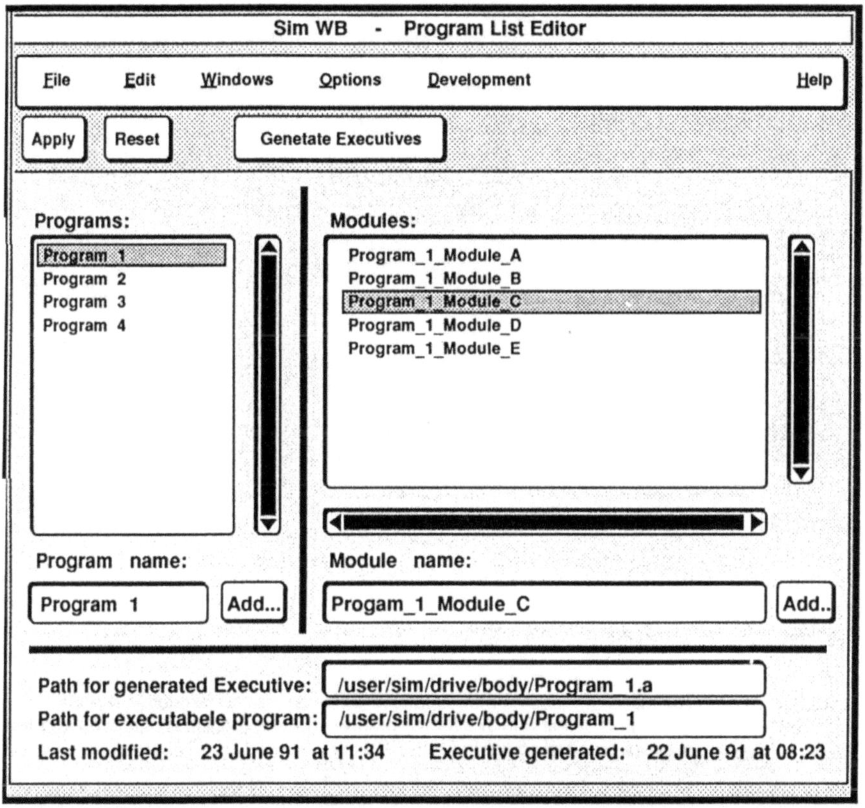

Registration /Maintanence of Simulation Worbench Applications

The complete GUI is programmed on the basis of X-Window and OSF/Motif. So instead of having a long description on these windows it is better and more easy to show some of these almost self explained screen layouts (windows).

7. The Solution, ... and the sun shines bright at the end of the tunnel...

SIMulation Workbench is an user environment that automates and facilitates the process of monitoring and control complex simulation environments with its graphical user interface. SIMulation Workbench is an easy to use tool for the administration of such software tasks and one of the first CASE tools specially designed for the simulation environment at all.

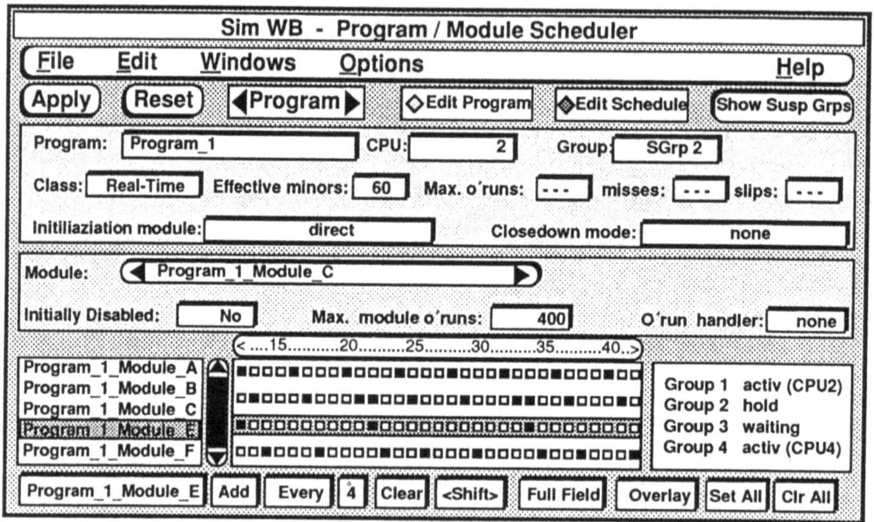

Simulation Workbench - Controlling the Application Schedule

Appendix:

A. Terms and Definitions

Major period	Time span of a major cycle.
Program	An application is made up of one or more programs. Each program is a collection of modules (see below) and the associated executive modules generated by the Sim WB for the program. Each program is compiled and linked to produce an executable file.
Scheduled program	A scheduled execution instance of a program (a RTU process started from the program).A single program can be scheduled more than once. Each scheduled program is unique and consists of an application defined sequence of execution of some (or all) of the modules defined in the program.
Module	A subprogram unit within a program that is registered to be called by the cyclic executive. This can be an Ada subprogram, a C function, a FORTRAN subroutine, or an assembly routine.
Schedule module	An application defined scheduled invocation of a module within a program.A scheduled program consists of a definition of priörities,frequencies of a set of scheduled modules.
Real-Time	For the purpose of Sim WB, a scheduled program is termed "real-time" if overrun detection and cycle timing of its scheduled modules is called for at runtime. Scheduled modules within a scheduled program, which runs "Real-Time", need to be invoked at application specific frequencies.

Background	A scheduled program is termed "background" if it is meant to be run AFTER all scheduled "Real-Time" programs have finished in a minor cycle. No overrun detection nor cycle timing is performed for scheduled modules in a scheduled "background" program. These generally consist of house-keeping processing tasks. They simply get preempted and resumed later when idle time is available on the processor.
Executive	The main routine of an application program responsible for interfacing with the SPS to effect user level subprogram unit scheduling. This code is automatically generated by Sim WB and is linked in by the user with his application code before the simulation starts.
Configuration	In the context of Sim WB, the configuration of an application refers to the set of specifications for parameters describing fully the characteristics of the frequency-based application. This includes the programs that make up the application, the characteristics of the scheduled programs and scheduled modules within them, etc.
Scheduling group	A scheduling group is a logical entity consisting of one or more processors, on which a number of programs are scheduled. All scheduled programs in a scheduling group are driven at a common base frequency.
Effective frequency	A scheduled program can be driven at the base frequency of the scheduling group that it belongs to, or at an effective frequency which is a factor of the base frequency. For example, if the base frequency of a scheduling group is 60 Hz, and the effective frequency of a scheduled program is 30 Hz , then the scheduled modules within that scheduled program running at maximum rate will be scheduled at 30 Hz and they will have a maximum of 1/60 of a second to perform their designated tasks.
Standard frequency	A scheduling group can be programmed at one of the predefined standard frequencies,, such as 60 Hz or 30 Hz, which implies a major period of 1 second and a number of minor cycles that equals the frequency.
Custom frequency	A scheduling group can be defined to have a base frequency which is not integral by specifying the major period. For example, a major period of 2 seconds and 15 minor cycles per major cycle will give a custom frequency of 7,5 Hz.
Overrun	A scheduled module is said too have overrun if it is still active when the epoch of a minor cycle arrives. If a scheduled module overruns, its execution is terminated, and control of the scheduled program is returned to the executive. If a scheduled module overrun count exceeds an application defined value, the executive disables the scheduled module.
AST	Asynchronous System Traps supported by RTU.

B. Literature

All named literature was published by or within Concurrent Computer Corporation, Tinton Falls, New Jersey, USA.
This Paper was partially taken from the named literature [1 - 3]:

[1] Updated functional specification of SIMulation Workbench, by Charles Fung (internal paper)
[2] SIMulation Workbench Graphical User Interface (GUI) Functional Specification, by Jem Treadwell, (internal paper)
[3] Synthetic Periodic Scheduler, Product Data Sheet, MLDS 2802, 11/90,
[4] Understanding Real-Time-UNIX,Background Information Brochure,MLMIS 1383,

Widening Numerical Intensive Computing (NIC) Perspective – Shorten the Development Cycle

Armand HERSCOVICI

IBM France Scientific Marketing

INTRODUCTION

Numerical intensive computing (NIC) is born in the 1970ies, with the arrival of supercomputers. NIC had been used before (with IBM being an active partner). But the dramatic oil situation during the 1970ies, generating a worlwide economic crisis, settled a situation driving to an extremely fast development of NIC. Competition became much stronger than before, and companies had to develop new strategies to meet it. A fundamental trend in these new strategies consists in all the component enabling companies to be highly reactive to any competitor initiative. A way to be reactive consists in issuing as soon as possible (sooner than the competitor if possible) new performing products (more performing than the competitor's). Meeting this requirement implies two items, which end in making NIC a strategic weapon in the economic war:

- shorten new products development cycle
- optimise new products as much as possible

Reaching these objectives means developping new methodologies (which cycle components have to be acted upon, how should this be done, how?), and new organisations to implement these methodologies (how to organize people involved in product development).

Taking into account these new components results in widening more and more the understanding of NIC content. This new approach of NIC has been understood not very long ago.

It is interesting to note that the understanding of the way a whole company works has undergone a similar process («systemic approach», as this will be explained below). As a matter of fact, the widened approach of NIC is an homothety, at R&D level, of the whole company global systemic approach.

Of course, the widened NIC approach applies fully to the world of Research. One could even say that Research is the motor in this evolution. Using supercomputers implies new techniques in term of programming and numerical analysis. Availability of huge computing power enables the development of totally new applications. Implementing these new characteristics requires intellectual ressources available in Research organisations only.

As a matter of fact, advanced NIC applications and corresponding techniques generally appear in large laboratories. They progressively invade the scientific community. Later, they enter industry, starting with high-tech (like aeronautical or space). They reach other industrial segments when the techniques have been «industrialized», which means that they have become «easy to use» by professionals working far from the world of Research.

TRADITIONAL PERSPECTIVE OF NIC

BIRTH OF SUPERCOMPUTERS

One of the first NIC application consisted in implementing numerical wind tunnel. This consists in mathematically simulating aerodynamic phenomenons around an object (say a plane). One can show easily that due to the algorithms used, and to the level of accuracy requested by engineers, approximatively 10^{13} floating point operations have to be performed. If this calculation is run on a 10 megaflops computer (which means the computer speed is 10^{7} floating point operations per second), it will last $10^{13}/10^{7}$ seconds, or 12 days (CPU time). Considering that:

- this amount of calculation is mandatory as products have to be highly optimise
- a 12 days response time does not contribute to shorten the development cycle time
- fastest scalar computers hardly run beyond 10 megaflops, either IBM or anybody else

(remenber this explanation (takes place at birth of supercomputing)

One have to admit that technology is unable to provide the requested power. As a consequence, computer vendors have to compensate technology by «tricks» in architecture. This is done by detecting which are the most frequently used functions in the calculations, and implementing an optimised hardware for these functions in the computer. We then get a specialized computer, namely a supercomputer, which has an extremely good price/performance ratio when running the hardware implemented functions (but a very bad one when running anything else).

APPLICATIONS

A supercomputer is a specialized computer enhencing the speed when running a particular algorithmic structure: the sequence of operations Multiply and Add within a loop (we still stand in the 1970ies). Only those applications implementing that structure frequently enough can run efficiently on a supercomputer. It happens that many NIC applications meet this requirement.

A large categorie of NIC applications deals with the simulation of continuous fiels: fluid dynamic, structural analysis, electromagnetism, etc..The physical laws which

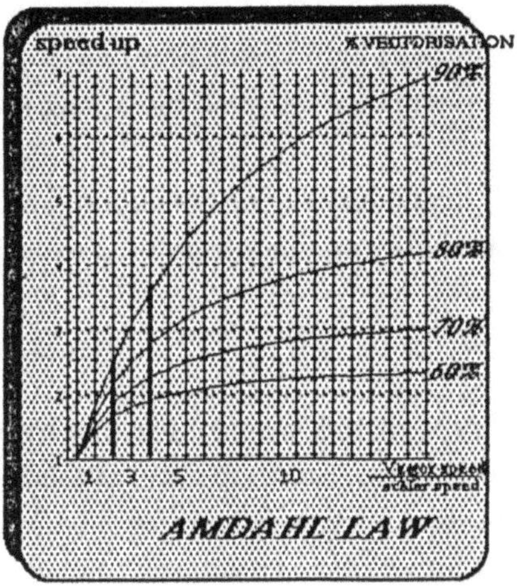

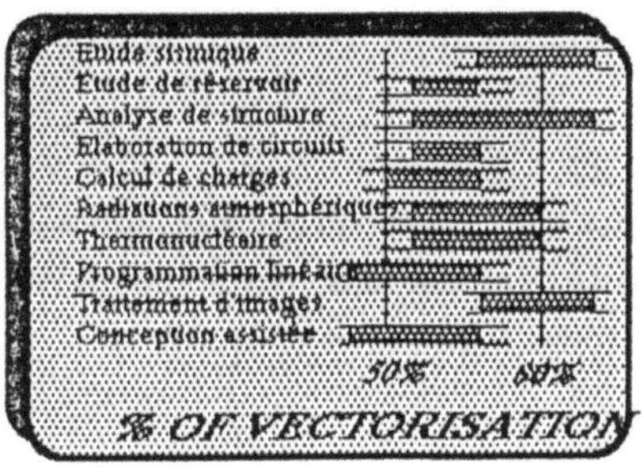

govern the corresponding phenomenons are all expressed mathematically as sets of partial differential equations (PDE). PDE are solved numerically by discretisation of the field, which leads to transform the set of PDE in a set of linear equations (if the set is not linear, one know how to make them linear). Linear systems are solved by matrix operations. It just occurs that within the algorithms implementing these operations, the loop Multiply -Add is to be met very frequently. As a consequence, supercomputers are fast when dealing with continuous fields.

Some other domains implement the same kind of algorithm, like signal processing.

Continuous fields simulation, signal processing, and some others, may be run efficiently on vector supercomputers.

LEGENDS VERSUS REALITIES

Legend

Vector processors are specialized in running the loop «multiply-add». A scalar (non-vector neither parallel) processor needs generally between 6 to 8 machine cycles to perform that loop. Due to a special way of implementing parallelism, a vector processor performs the same loop in one cycle, which means two operations per cycle (one floating point multiply and one floating point add). Given the cycle time, one can easily derive the performance in term of number of floating point operations -flops- per second (as a matter of fact, the unit is megaflops).

With this approach, a computer with a 15 nanoseconds -ns- cycle time has a performance of 133 megaflops -Mflops-. Another one with a cycle time of 4 ns would lead to a performance of 500 Mflops. This performance is called the «peak rate».

Too many users are still fascinated by these figures, considering the supercomputer at this elementary level. Of course, the peak rate performance has nothing to do with reality.

Reality

If the computer was really running at peak rate speed, this would mean that from its birth to its death, it would have performed the multiply-add loop only, as this is the only algorithmic structure where the computer executes two operations per cycle. Clearly, if our basic loop is met frequently, a lot of other structures (called non-vectorisable structures) appear in any scientific program. This includes any non-loop organisation, but also loops which hide «inhibitors» to vectorisation.

A non-vectorisable structure has to be run on a scalar processor, and a so-said «vector processor» always implements both a true vector processor and a scalar processor. The performance is then a composite of scalar and vector speed. We will see below that a lot of other parameters are to be taken into account.

The Amdahm law

The Amdahl law shows the speed-up which can be obtained from a vector processor, given its vector speed, its scalar speed (or the ratio vector/scalar speed), and the percentage of vectorisation in the application. Clearly, a significative speed-up can be obtained only if the percentage of vectorisation is high (at 50% of vectorisation, a vector processor with infinite vector speed would lead to a speed-up of two only). A major consequence is that a reasonable price/performance is achieved only for application with a very high level of vectorisation.

The figure below shows that a major part of applications are vectorised up to 50 to 80%. IBM has taken this fact into account, implementing a vector facility with a ratio to scalar speed driving to an average overall speed-up of 150% to 300% (which means + 50% to +200% above the scalar speed). The cost for this additive speed is only between 5 to 8% of the basic scalar machine.

A first step

Even though taking the scalar speed into account brings the peak rate down to a more realistic value, a lot of the computer parameters have still been omitted. This addresses the traditional performance components, like the storage, the channels, the disk drives, the operating system, the compilers, etc. Even though we will not discuss them, as they are not part of the subject of this presentation, they should never be neglected. To often, they are not considered in the benchmarks proposed by the users.

But up to now, only DP questions have been discussed. The initial objevtive has just been forgotten: the purpose is not to compute as fast as possible, but to shorten the development cycle. We still have to widely broaden NIC perspective to reach our target.

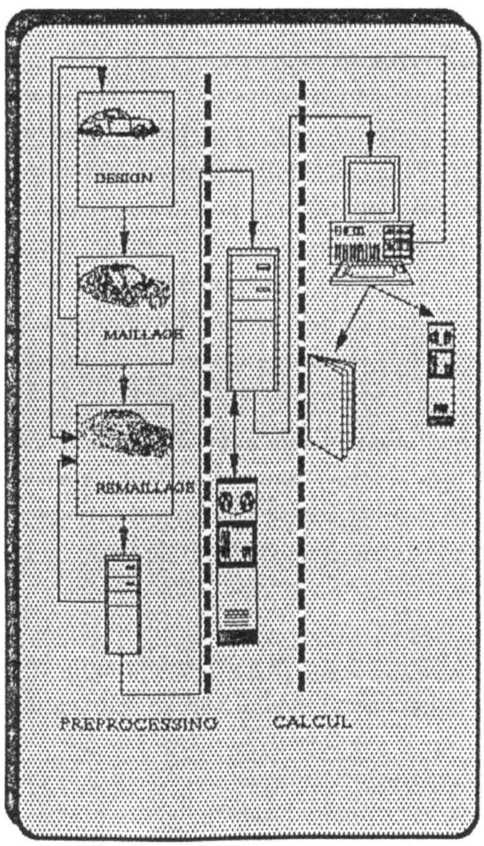

TOWARDS A WIDENED NIC PERSPECTIVE

A widened NIC perspective implies taking into account all the items which are part of the development cycle, keeping in mind that the only objective is to shorten this cycle. Quality in the project will appear as a consequence of the shortening methods. The explanation of the concept will be given using an example: the development cycle of a car body.

CHAIN OF APPLICATIONS

First step (linear analysis)

A first tentative shape has been designed by the CAD organisation. The calculation process starts when one wants to make sure that the body will resist to the different stresses. This is done using the finite elements methods. Calculations generally show that some reinforcement has to be provided in some locations of the structure, which drives back to the CAD process where the requested modifications are performed. The mesh is modified, a new calculation is run, eventually detecting new changes to be done, leading to new modifications in the mesh. This is an iterative process, at the end of which the body is correctly designed for that kind of stresses

Assuming the different steps in this approach (CAD, meshing, calculating) are optimized, the more the data flow between the steps run easily, the more the overall process is efficient. Becoming concious of this reality gives a first hint of how important is the **concept of integration**, together with the major role of the **quality of the links.**

This step generally does not requires a supercomputer. Between 3 to 5 months of an engineer may be necessary to design and modify the mesh.

Crash analysis

This application is a very demonstrative example of NIC dramatic growth in industry. Calculations take place in the non-linear domain.

Designing the mesh

In a frontal crash, the frontal part of the car absorbs most part of the energy, when the rear part undergoes a very low amount of stresses. The mesh has to reflect this reality, being very tight on the frontal part of the car, and becoming wider and wider towards the rear part. This means the mesh is different from the one used in the linear analysis.

This new mesh can be developped from scratch, requesting again three to five months which have to be added to the overall development cycle time. Another possibility is to take advantage of an eventual technical data base, where a former release of the mesh has been saved. Then the process consist only in modifying the original mesh, which may request a few weeks instead of a few months before. Such a data base would clearly save a few months on the overall development cycle time. Increasing the speed of the supercomputer up to the infinite (with a probable infinite associated cost) would not provide such a shortened duration.

The data base should allow to find easily not only the former release of the mesh (which one, for which project, etc..), but also all the items associated to it.

Designing the mesh for the crash, and testing it before launching the calculation may require in the rage of one month of engineer, and five to ten hours of supercomputer.

Calculating

Assuming the implicit method, the calculation is run through a sequence of some 1000 to 10000 steps, each step taking the body at «t» time and projecting its status at «t+dt». Each step is equivalent to a linear analysis, in term of computer ressources. A linear analysis may produce a 5cm thick computer print out. If the whole crash computer print out had to be issued, this would mean printing a document 1000 to 10000 times 5cm thick, or 50 to 500 meters thick.

Clearly, non linear calculations introduce a totally new dimension in NIC, in term of huge amounts of data. This drives to two mandatory items in the NIC system:

- a highly efficient system in term of I/O handling
- a data base (again!) where results may be saved for future postprocessing.

Postprocessing

The end user has now to deal with a huge amount of data. The only way not to get lost in this step is to use graphical display (either workstation or graphical system connected on the mainframe), which will allow synthesis, etc. This alowws fundamentally to show what is invisible (stresses, intermediate status of the structure, etc), and to include the human eye, with its capabilities of sorting, synthetising, understanding, in the development process.

Large quantities of information are transfered between the data base and the graphical display. Here again the whole process is efficient only if links (interfaces, networks) allow a smooth and large transfer of data. No supercomputer is requested in this step.

Finally, a report is issued, using some text processing or in-house publishing system.

The development cycle starts at linear analysis steps, and ends when the report is issued. This is what a company should aim at shortening. This is much wider than just trying to shorten the traditional NIC step, and implies an enlarged perspective of NIC.

The Chain of applications as a system

The table below sumerizes the time spent on the differents steps of the development cycle.

Clearly, a car body project requests some 3 to 5 hours of supercomputer per month per engineer. Emphasizing traditional NIC performance means taking care of 3 to 5 hours among the 180 hours an engineer is working in a month.

Shortening the development cycle implies going much beyond this restricted target. As explained above, significant contributors will be:

- a coherent step-link system. Data will travel easily without delay. This is the integration concept, which implies quality interfaces, and cooperative processing. If integration is implemented, together with an adequate technical data base, the development process becomes «industrialised». Many successive releases may be worked upon, with a high level of quality in the process (the data base will allow an efficient management), driving to a highly optimized product.

- a balanced system. What would be the use of spending most part of one's budget on an extremely fast supercomputer, if the huge amounts of data (the more powerful the computer, the larger the amount of data) could not be easily transfered or handled.

In a top-down approach, the chain of application has to be perceived as a system, made out of different elements. These elements are not independant from each other, they are linked together. As a consequence, the optimum of the system is not reached by optimizing each one of the elements, as links have to be taken into account in the optimizing process.

This analysis can be considered as the «homotetic», at R&D level, of the modern approach of the overall company: a company is no longer seen as a set of functional units where each unit has to be optimized. Links between units are taken into account, and very often appear as the bottleneck in the system.

CONCLUSION

The fundamental objective of a company in the R&D process, when using a NIC organisation, is to shorten the development cycle. This leads to a widened approach of NIC, which implies a global understanding of the environment, and a systemic approach of NIC centers.

At the lowest level, the vector speed of the supercomputer does not appear as a major contributor. Balancing vector/scalar speed according to applications characteristics enables a good price/performance.

Other DP parameters contribute to the system performance with an equal level of importance: storage, disks, I/O system, operating system, compiler have to be considered in a balanced perspective.

At R&D department level, the systemic approach is the right one: this means technical data base (a safe for the company where the know-how gets richer and richer), applications integration (quality interfaces and links), cooperative processing with powerfull workstations and market standards systems.

The IBM strategy towards R&D organisations is to propose solutions for all the items in the widened NIC approach. The available product line meets this objective, including the capability to work with non-IBM systems.

Supercomputers and Hardware Tools

Engineering Simulation on High-End Computers

J. Pareti

Digital Equipment Co.
Manufacturing Industries
Freischuetzstr. 91, D-8000 Munich 81
Germany
Telephone [49](89)9591-1093

Summary

Engineering simulation in the automotive industry is an essential part of product development and it encompasses applications in such disciplines as Finite Elements Analysis, Crash Analysis and Computational Fluid Dynamics (CFD) which allow the designers and analysts to try-out in the computer different variants to meet product performance and reliability objectives: all this long before prototyping. Having expressed the governing differential equations for the engineering problem, numerical techniques are employed which typically lead to iterated matrix operations to compute such parameters as structural stresses, temperatures and flow properties. Usage of high performance computers makes it possible to realistically examine real problems with acceptable response times. Price / performance ratio is mostly a key consideration when acquiring a computer system. Vector processing is relevant for problems with a large number of degrees of freedom. Scalar performance prevails in I/O intensive applications and with significant parts that are not vectorizable. A suite of software tools to efficiently compile and debug source codes and to detect bottle-necks at run-time is part of the developing environment. Human expertise in processors technology, engineering applications and numerical modeling is a decisive factor to meet the users' requirements using team work. Finally, standard compliant open systems are a major requirement of the users for the best match application-platform. In this respect the system vendor's commitment to open systems is a key consideration.

Product development in the Automotive Industry

Reduced time to market is a key success factor in the automotive business. Japanese manufacturers take an average of 4 years from the conceptual design of a new car to the showroom, vs. 5 years for North American manufacturers. The highly competitive market commands R&D expenses which are in the order of 5% of total sales revenues, which cover product development costs from engineering to manufacturing and servicing. In particular, Computer Aided Engineering (CAE) is a key technology area enabling faster prototyping and ultimately reduced time to market. CAE tools enable analysts and designers to perform a "what-if" analysis in the computer to discard a number of inferior designs long before prototyping. CAE applications require graphic tools and interfaces to convey design data to the analysis modules such as Finite Elements Analysis and Computational Fluid Dynamics. FEA is employed to solve problems, such as component strength optimisation based on linear elasticity stresses, complex non-linear problems, e.g. including plasticity and residual stresses, and dynamic analyses on assemblies such as motor-block and gear-box unit, for designing out noise. FEA is also used for crash analysis, air-bags design and reinforced bumpers design for safety. Computational Fluid Dynamics is employed to analyse combustion processes and in general any fluid-dynamics related process such as gas flow in a catalytic converter by calculation of velocities, pressure, turbulence and temperature which will lead to shape optimisation for performance.

CAE applications

Design in the form of CAD data is the starting point for the analysis. Automobile body design is usually performed using surface modeling (e.g. Coon's and Bezier's patches, and more recently NURBS). The model is either passed on to a FEM pre-post-processor within the same CAD system or exported to an external one using e.g. IGES or VDAFS. Based on this the analyst makes model simplifications such as removing features which are inconsequential for analysis and uses the pre-processor to generate nodes and elements. Finally he includes boundary conditions to isolate a component from an assembly, still retaining the functional relationship with the assembly. Boundary conditions are typically in the form of restrained points or nodes, loads, as well as constraints of nodes obeying a mathematical relationship. Depending on the required results the problem size will increase, for example to detect stress concentrations in the surroundings of irregularities such as holes, notches, etc. This will result into a less stress-sensitive component. Within dynamic analysis one can reduce the level of vibrations (and noise) transmitted to the interior of an automobile for various operating conditions. A first step to this is achieved by performing a modal analysis to identify the system eigen-frequencies and related modes of vibrations based on 3D models. This may result in a model of tens of thousands of nodes, requiring in the order of several hours computing time and producing several GBytes of output data. As the design-analysis sequence is by nature interactive the designer would want an almost immediate response from the computer. Even if a complex problem is solved using hired CPU on remote supercomputers in less than one hour, long turn-around times will often delay the design-optimisation chain. This situation will confront the enterprise with the need of identifying applications which should be offloaded from an external supercomputer by employing an in-house dedicated facility based on a multi-user distributed environment at competitive price/performance levels. The benefits are reduced cash costs for external CPU, increased turnaround time and hence speed-up of the "what-if" analysis for an entire class of engineering problems up to thousands of simultaneous equations solution.

Finite Elements Analysis

Finite Elements Modeling converts a structure into a set of elements connected at nodes, some having loads and/or restrained degrees of freedom. Each element deforms so that nodal reaction forces arise depending on the nodal displacements and on the element stiffness matrix $\mathbf{K_e}$. Within the whole structure each node undergoes reaction forces from the elements to which it belongs and possibly external forces, the vector sum of which must, for equilibrium, be zero. This gives one scalar equation per node per degree-of-freedom. For the whole structure the following matrix equation applies:

$$\mathbf{K}\,\mathbf{X} = \mathbf{f}$$

where $\mathbf{K}$ is the global stiffness matrix, $\mathbf{X}$ is the nodal displacement vector, which is to a large extent unknown, i.e. with the exception of restrained nodes, and $\mathbf{f}$ is the applied load vector. The coefficients of $\mathbf{K}$ can be thought as submatrices of influence numbers for all the nodes of each element containing the generic node i at which force equilibrium is considered. Clearly all elements containing i contribute to the stiffness and consequently the submatrices of $\mathbf{K}$ are determined by summation of the appropriate element stiffness coefficients. The nodes of those elements containing i should ideally be indexed in such a way as to minimize the differences relative to i, in order to achieve a bandwidth of $\mathbf{K}$ as narrow as possible and hence to minimize the solution time. Nodes re-indexing is generally performed by FEM codes. A solution algorithm is based on the triangular decomposition of $\mathbf{K}$ into the product of a lower and an upper triangular matrix, or $\mathbf{K}=\mathbf{LU}$, followed by forward elimination and back-substitution. A more general problem is how a component behaves, which can be considered as a continuum delimited by boundary surfaces. The solution method is analogous to the discrete structure solution above, but approximations have to be made to (a) subdivide the continuum into finite elements to model the part, and (b) to numerically compute the stiffness matrix. The elements shape influences how the part is approximated; these are defined in (curved) local coordinates which are then transformed to a global system. To compute the stiffness matrix the virtual work equation is employed: (i) displacements are expressed as a function of nodal displacements and "shape functions", (ii) strains are obtained by differentiation of displacements and (iii) stresses are obtained from strains and a material constitutive model. The external work, product of nodal loads and virtual displacements is equal to the internal work of the virtual strains and of the stresses induced by the real nodal displacements. Step (i) involves an approximation of the displacement by linear, parabolic or cubic polynomials, or "shape functions", based on the nodal displacements. When local/global co-ordinate transformation and shape functions (e.g. for displacements) use the some supporting nodes the element is called isoparametric. Both mesh finesse and the polynomial degree affect the rate of convergence. High stress gradients call for locally dense meshes. Moreover as too rapid transitions from coarse to fine mesh impair the results accuracy due to distortions one should rather prefer multiple points constraints, or relations among displacements by which the effective number of unknowns (in the low gradients region) is reduced. The polynomial degree affects the way the function approaches the true value; it is useful to think in terms of Taylor's series : for example if linear elastic elements of side lengths h are employed the displacement error is $O(h^2)$ but the stress error, given that the stress is constant over the element, is larger, or $O(h)$. For this reason less higher order elements may be preferable. Another approximation is introduced in evaluating the strain energy (integral of the product of stresses and strains matrix) to obtain the element stiffness matrix by means of numerical quadrature : this is based on special points in the element, the position and number of which is optimized for convergence in regular elements (reduced integration). The finite element method is also applied to other Engineering problems than stress analysis using an integral form of the governing differential equations (e.g. for heat-transfer, modal analysis, field problems). The differential equations are first expressed in vector form then multiplied times arbitrary vectors, or weight functions. The product is integrated over the domain delimited by the boundaries; on those the functional conditions are similarly treated. This process requires usage of continuous shape functions: for example in the displacements problem C1 continuity of the shape functions is normally required for convergence of the integral equation (i.e. continuous strain which is the derivative of displacement). This is called the "strong form" of the integral equation. However even if C1 continuity is violated convergence may still occur for elements that pass the "patch-test" on a number or patch of infinitesimal elements (similar to the given ones) on which for a given strain the correct stress is computed . To achieve convergence for a wider class of problems an integration technique can be applied giving the "weak form" by which C0 continuity of the shape functions will be sufficient as long as the weight vector is at least C1 continuous. Commercial FEM packages are equipped with various morphologically different elements to model most engineering problems. Some common ones are bars and springs (carry load along the axis), beams (suitable to model bending of shafts, spindles, etc. are based on St.Venant's theory), plates for 2D plane-stress problems. For 3D problems: thin shells (differ from plates as they take a load normal to the plane; the in-plane and out-of-plane deformations are however uncoupled); axis-symmetric elements (suitable to model turbine disks, pressure vessels, plain bearings, etc.), bricks and other solid elements, for general 3D problems. Gap elements are used to model contact problems, whereby elements penetration is not allowed.

Non linear analyses

Crash analysis involves both geometry and material non-linearities due to on one side large deformations-large strain (e.g. buckling) and material permanent deformations and residual stress on the other side. Results are typically stored at several points during the impact and will comprise displacements, velocities, accelerations, stresses and strains. As an example the car front part is accurately modeled, taking into account engine and engine supports, while the rear part may just consist of beams. Traditionally the FEM results have been validated with experimental data; crash analysis proves however to be even more far reaching, as it allows to understand failure modes of hidden parts of the car assembly for which experimental observations are difficult to obtain. The simulation results are input to the test team as to where sensors accelerometers, etc. should be placed. For a model of 10000 elements, about 10 Mbytes of information will be stored. Another application area is metal forming (sheet metal and massive forging process) which involves material and geometry non-linearities, as well as friction at the tool-part interface (where gap elements can be used to control the sheet position) and shell thickness variations during the process.

Dynamic analysis

Modal analysis is used to design out noise and vibrations in a mechanical system. When the design cannot be changed allowable operating conditions can be specified. (e.g. speed or speed ranges for a rotor). To perform modal analysis the majority of elements can be employed as in static analysis; in addition the elements inertia properties to form the dynamic equations have to be specified. For a structure, depending on inertia and stiffness the eigen-frequencies are determined. Next the mode shapes are identified, or preferential deformation fields associated with a given eigen-frequency, such as for example the torsional modes of a shaft,etc. This gives the free vibrations response. One decomposes then the applied load into a sum of harmonic terms with different amplitudes and frequency e.g. using Fourier transform for the torque function in a piston engine. If one of the excitation frequencies coincides with one eigen-frequency resonance occurs for that frequency. The stress amplitude at resonance depends on the product of system gain (amplification) and input load amplitude. At resonance the stress reaches theoretically infinite values; in practice due to the amplification depends on frequency and system damping, some harmonics may be resonant, but without significant effect. This gives the forced vibrations response. Noise is associated with vibrations in mechanical systems which are calculated. One could define the sub-system imposing a forcing function to the structure (e.g. driveline as the sub-system; car frame as the structure) and then establish the frame accelerations due to drive on a given road. Experimental work associates observed noise levels to measured acceleration levels. Another application is rotodynamics (spindles, turbine shafts, gear transmissions, crankshafts) .

Computational Fluid Dynamics

CFD is a key application within automotive engineering for components optimisations, such as combustion chambers, profiles, etc. The starting point is typically CAD surface data. A finite volume technique can be applied to the surface bounded domain, giving elements, on which conservation of energy, mass and momentum for the fluid is applied. Combustion is modeled by introducing additional terms, or sources to describe phase change (from liquid to vapor) and chemical reactions resulting into a different distribution of physical properties. The partial differential equations are approximated by finite differences giving algebraic equations to be solved iteratively. Velocities, pressure, temperature distribution as well as the phase composition as a function of time are computed. The equations solver consumes 50 to 80 % of the CPU time for problems in the order of 30 to 50 thousands cells. Vector processing speeds-up the equations solution for algorithms without recursions or data dependences which would inhibit vectorisation. For example the red-black or chessboard computational scheme in the Gauss-Seidel iteration technique for linear equations systems and a sparse coefficients matrix gives vector-to-scalar speed-up factors from 2 to 5 depending on the computing architecture. An example is the CFD code FIRE of AVL. Iterative methods are preferred to direct ones to save memory and direct methods (e.g. "skyline") are confined to smaller problems.

High-end computing technology overview

Performance is achieved through the computing architecture. It can be measured using standard bench-marks, as Perfect Club consisting of engineering applications. Global performance is determined by scalar performance, memory bandwidth, i.e. the rate of data transfer in and out of memory, vector processors performance, which is measured in MFLOPS (millions of mathematical operations, floating-point operations per second), number of processors operating in parallel, and I/O devices access speed. Fast access cache-memory may be implemented in addition to lower cost main memory and be efficiently used through compilers. High bandwidth allows to entirely convert processors technology into performance. A balanced architecture suits best a multi-user environment that one can simulate in a multi-stream bench mark. The figure below shows the performance ratings of various vector and scalar computers running the FEA code ANSYS

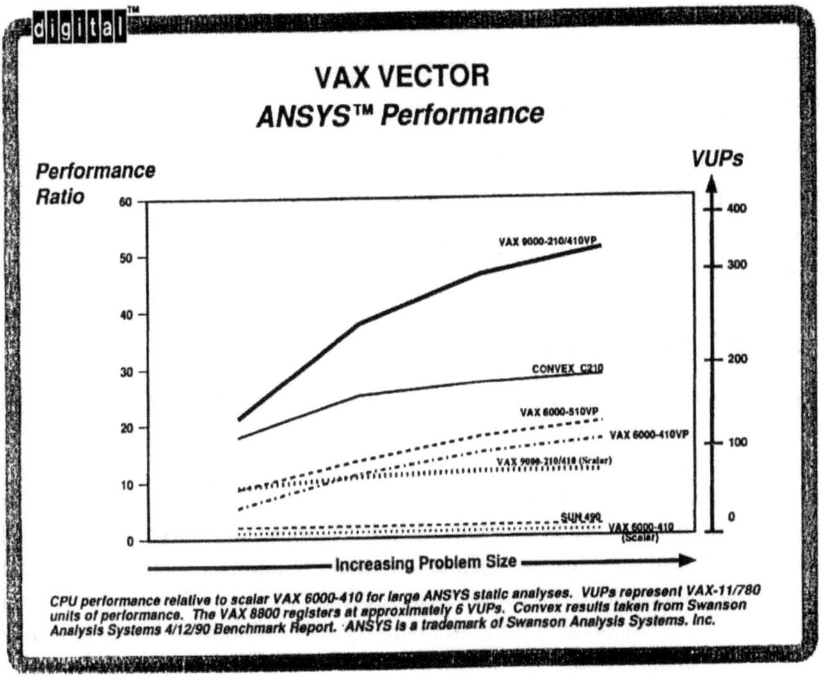

Client/Server

Efficient use of resources can be achieved by distributing the computing power and finding the best match application-platform. The client/server environment provides common resources to a users' community and supports efficient methods for data and system management. The benefits are in data availability, reliability, accessibility and security. A high performance computer can be employed for processing numerically intensive applications in batch and also as multi platforms server.

Open Systems

A key requirement of the user community is the ability to run an application on any suitable platform by ensuring compliance to standards such as :

- OSF / MOTIF and DCE
- IEEE POSIX
- ISO-OSI, SQL Languages, ODA / ODIF
- X/OPEN- CAE, XPG3, Branding

POSIX is a standard designed by IEEE to meet the definition of Open Systems and it is endorsed by organizations such as Open Software Foundation, X/OPEN, UNIX International, Uniforum, European Commission, NIST and ISO.

MIPS and MFLOPS : when and where

Vector and parallel processors make applications run faster that are numerically intensive (floating point & integer) and that repeatedly do the same calculation on a series of data, such as in matrix algebra operations. The MFLOPS rate depends on the individual processors, pipelined operations, and processors coordination for parallel execution on a same piece of code. Scalar computers operate on a single datum at a time and are rated by the millions of instructions per second (MIPS). Scalar performance is important for I/O intensive applications and where data dependences prevent vectorisation and parallelisation. Scalar computers are also referred to as Single Instruction Single Data (SISD). Vector computers are also called Single Instruction Multiple Data (SIMD). Multiple Instruction Multiple Data (MIMD) refers to Parallel and Massive Parallel Processors.

Vector processing

Vector processing can be compared to an assembly line in which an operation is broken-down into a number of sub-tasks. For example and ADD operation may consist of three sub-operations. When the following FORTRAN statements are processed in scalar mode:

```
      DO 10 J=1,N
   10 X(J) = A(J)+B(J)
```

it will take three cycles to process the ADD operation for ONE result, e.g. X(1)=A(1)+B(1). When the same statement is translated into a vector instruction, the input data (A and B in the example) will be loaded into vector registers. The individual sub-tasks of ADD will operate on different array elements. After a three-cycle latency required to fill the pipe there is one result per cycle. If the vectors are e.g. 30 elements long it will take 32 cycles to process the entire ADD, whilst a scalar processor would take 90 cycles. The speed-up factor is in this case 90/32=2.8. The figure below is a graphical representation of the concept for the add example and it shows that after the latency period one result is produced each cycle.

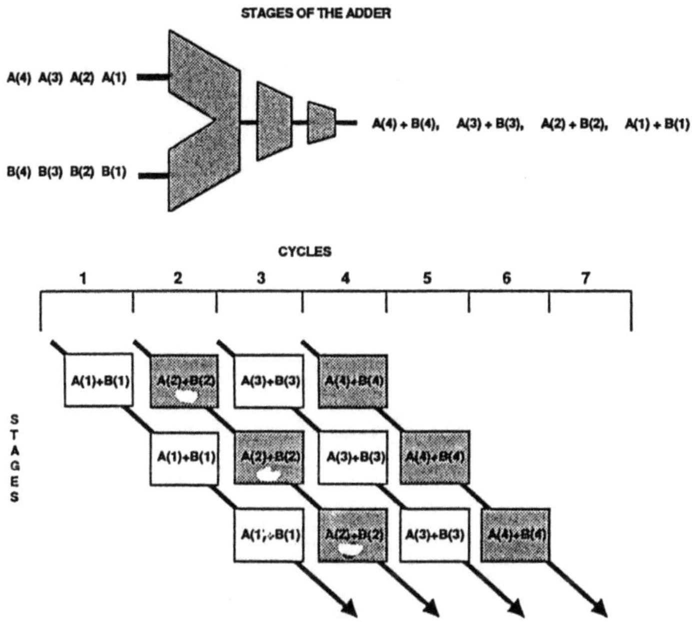

Types of vector processors

Attached array processors are loosely coupled vector co-processors that hang off the I/O bus. They usually have their own instructions set and memory and different data formats compared to the scalar processor and different compilers are required producing each an object code to be linked into one executable file. I/O bandwidth limits the continuous performance of an attached processor; it is therefore required that vector operations greatly outnumber I/O operations. In contrast, ***integrated vector processors*** are tightly coupled vector co-processors and are used in most high-end computers. The scalar processor is designed to interface the vector processor using a high-speed connection; both processors share main memory, data formats, cache, translation buffers and other CPU states. The main advantages over loosely-coupled processors are for problems with a high ratio of data per vector operation and in the software development environment being consistent through common control by the operating system and based on resources sharing. Another feature is whether data is operated directly upon from memory to the processor(s) (memory-to-memory), or if the data is stored in vector registers which are then operated upon by the functional units (register-to-register). A memory-to-memory architecture yields best results for sufficiently long vectors to compensate for the relative long latency in data access from memory. Usually memory-to-memory is implemented using high speed - high cost memory units.

VAX vector architecture

The VAX vector architecture consists of tightly coupled scalar and vector processors and it is a memory-to-register type. Data transfer in and out of memory has different implementation on the VAX6000 (through the XMI bus at 80 MB/sec) and in the VAX9000 (using a cross-board system configuration unit, that greatly enhances the bandwith to 1000MB/sec).

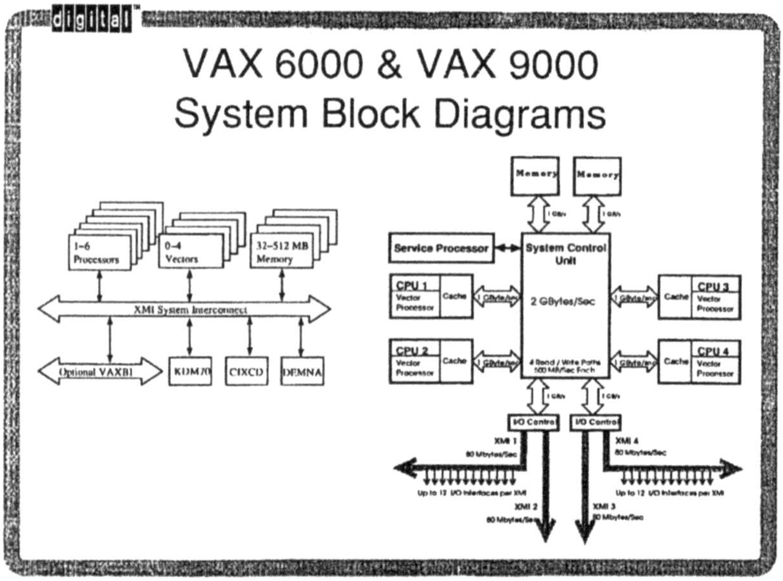

The memory system consists of high speed cache memory of limited size in addition to normal memory (e.g. the cache memory of the VAX9000 is 128 KB). Ideally all the data used in load and store instructions should be contained in cache; too large a cache would however be too expensive. Therefore optimum use of a relatively small cache is made possible by compilation, e.g. using array blocking techniques which will enable the processor to work on small parts of large arrays at a time. The vector processor consists of functional units operating on 16 vector registers, each of which having 64 cells of 64 bits each. The vector register size is an optimum balance to have as little as possible overhead when copying data to-and from memory into vector registers (this overhead is large for ***short*** vector registers) and at the same time have as little overhead as possible when transferring the registers content to cache at context switching in a multi-tasking environment. (This latter penalty increases for ***long*** vector registers). The fastest I/O devices are RAM disks having low application latency (1 msec), highest I/O request rate and bandwidth, which can store entire files, directories and devices. Volatility and large amount of memory required are practical limitations thereto. Solid State Disk have application latency of 2.4 msec, are ideal for write intensive files and have I/O request rates 10* magnetic disks. Finally conventional magnetic disks represent the lowest cost solution. Special software binding several disks to one set which will be accessed as one volume increases I/O throughput to 40 MB/sec and is recommended for I/O intensive applications both vectorised and scalar.

Vectorizable code and data dependences

Data dependences may prevent vectorization and can be recognised when the same vector is used for both load and store operations within a loop and the required inputs are calculated along the execution of the loop *(recurrence).* The compiler can resolve some forms of data dependences , e.g. using additional vector registers for saving intermediate results. Recasting the algorithm may eliminate data dependences. In some cases this can be simply done, by e.g. splitting a loop into two loops for disjoint access of the vectors; in other cases revised or approximated formulas may be necessary. The compiler can also generate machine code based on a given condition whose validity can be asserted at run-time. Finally vectorisation can be forced using a compiler directive but the correctness of the results must be verified by the user.

Vector performance

An application will always contain a part of code which cannot be vectorised. This limits the global attainable speed-up depending on the vector-to-scalar speed-up factor. As an example, if 3/4 of an application can be vectorised the theoretical maximum *global* speed-up factor is 4 when the performance of the vector processor goes to infinity and hence all the time is consumed to process scalar instructions **(Amdahl's law).** The vector-to-scalar speed-up decreases for short vectors and there exists a critical length below which the overhead required for the vector processors to be used exceeds the benefits *(cross-over point).* In such cases it is possible to prevent vectorisation by compiler directive.

Software Tools

The compiler generates both scalar and vector instructions, resolves source code constructs to vectorise and accepts directives to force or suppress vectorisation. It also optimises cache memory usage. It then informs on vectorisation.
A symbolic debugger is a tool for program development to inspect, and/or alter val-

ues of selected variables at run-time. Special Editors combine normal editor functionalities with detection of ambiguous data dependencies. An analyzer is a tool to pinpoint execution bottlenecks in programs and measure the time individual modules, and statements within a module use when the program runs on a given input data set. Emulators are used to run a vectorised code on a non-vector machine at program development stage. High-performance mathematical subroutines for frequent computation-intensive subtasks are available, such as Basic Linear Algebra (BLAS) for vector and matrix operations and Signal processing .

Case Study

The following sample problem demonstrates the application of vector processing to Finite Elements Analysis. A FEM program taken from Zienkiewicz "The FEM method" has been compiled and linked with and without vectorisation. As a sample problem a plate clamped on one edge and with a point load on the opposite free edge has been modeled in 2D with 3000 nodes and in the order of 2900 linear quad elements for linear elastic analysis. The results agree with beam theory for a beam aspect ratio of 20; then more elements have been introduced to the model size indicated above giving an aspect ratio of 1.3. This model size is reckoned sufficiently large to demonstrate the benefits of vectorisation. The stiffness matrix is stored using a profile technique then decomposed into LU product. The displacement vector is computed by forward elimination and back-substitution. Using a performance coverage analysis tool it was possible to ascertain that over 90% CPU time is spent in the solver, having subtracted the CPU time for I/O. Preliminary results indicated in the order of 1 min. CPU time on a VAX 6000-530 and only a marginal speed-up factor from vectorisation which is ascribed to the recursive nature of the solver.

Conclusions

Numerically intensive Engineering simulations can be efficiently performed on vector and parallel processors. Understanding the application, methods, technologies makes an essential contribution to fully unleash high-end computing performance; this is achieved by team-work and dedicated resources for application porting and optimising e.g. within FEA and CFD. Finally, full commitment to open systems enables portability and allows Users to benefit from distributed computing and price/performance advantages.

Corrosion and Ecology

Dynamic Corrosion Testing at BMW

Dipl.-Ing. Heinz Mühlberger and Dipl.-Ing. (FH) Ulrich Stricker
Bayerische Motoren Werke AG

1. SUMMARY

The principal factors influencing corrosion on vehicles in the possession of customers are high atmospheric humidity, high temperature, airborne spray containing salt and frequent exposure to thawing agents and other anti-slip substances spread on the roads.

The test procedure consists primarily of the stationary, thermal and chemical exposure phase conducted in the climatic chamber, the dynamic chemical and mechanical load phase involving actual driving of the vehicle and the final latent incubation phase.

The relative value of the results is principally confirmed by comparison with a control vehicle subjected to the same treatment at the same time; results have proved very satisfactory for the ongoing development of manufacturing processes and vehicle design concepts.

2. INTRODUCTION

The resale value of older motor vehicles continues to depend to a decisive extent on the amount of body rusting. Owners of automobiles in the top price bracket are particularly critical of visible signs of rust on the car's outer panels, since although this rusting is purely "cosmetic" in character, it affects the car's appearance so that it no longer conveys the desired air of distinction and luxury.

For this reason, the avoidance of premature rusting is of considerable importance. The efficacy of new processes, materials and design measures in restricting corrosion therefore has to be confirmed by reliable testing methods related to actual vehicle operating practice before they can be approved for use on series-production vehicles.

The test procedure must comply in particular with the following main objectives:

- It must simulate the pattern of corrosion to which the customer's vehicle will be exposed, in particular in areas where operating conditions tend to encourage corrosion
- It must be greatly speeded up in order to keep development times short
- The results must be satisfactorily reproducible.

3. FACTORS INFLUENCING CORROSION ON MOTOR VEHICLES

3.1. Atmospheric burden

In the classic, natural atmosphere, that is to say without any artificial effective substances in the air, the main factors which promote corrosion are high temperatures and high relative humidities. These are further enhanced by substances suspended in the air. It is regarded as proven that in coastal areas, the salt in the atmosphere and the high relative atmospheric humidity lead to more marked corrosion than in continental climates. The occurrence at the same time of high air temperatures, for instance those encountered on the Indian Ocean along the east coast of South Africa, result in an aggressive atmosphere which is critical with regard to corrosion. Industrial areas and other major conurbations in industrial countries are also notable for an atmosphere in which the emission of suspended atmospheric solids from the incineration of fossil fuels and from other chemical and thermal processes associated with modern civilization imposes a serious burden.

It has also been proved that these substances which initiate corrosion are not disseminated by adhering to the water droplets in natural rainfall, but at a relative atmospheric humidity of more than 60 % are already attached to the suspended water molecules of the moisture vapour in the air. This situation explains the presence of conditions which particularly encourage corrosion in areas near the coast which also happen to have a high concentration of industry.

3.2. The effect of substances spread on the roads

The practice in various countries and regions of spreading substances on the roads to enhance tyre grip has a significant effect on the corrosion pattern exhibited by motor vehicles. This has been shown for example by the investigations of I. Kawamoto and K. Nishii in North America.

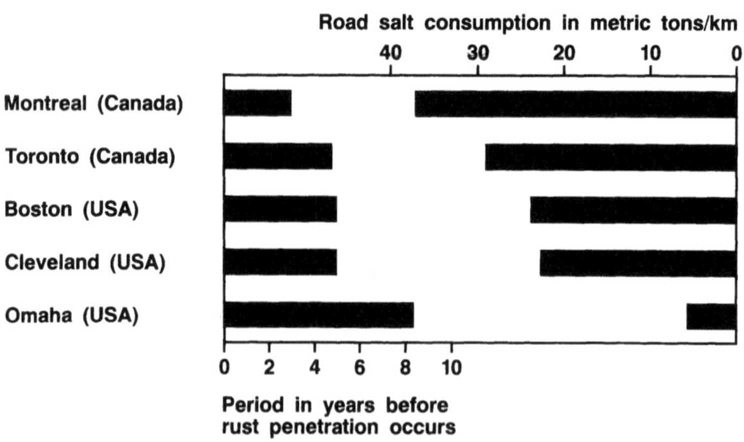

Fig. 1
Rust penetration in relation to volume of road salt used in selected North American cities [1]

The following trend in the spreading of such substances is identifiable: wet brines made up from a mixture of sodium chloride and calcium chloride are being used increasingly, together with other anti-slip substances such as stone chips.

According to statements issued by the Works Department of the Bavarian State Ministry for Internal Affairs, the use of salt in wet form has the advantages of being more easily metered when it is distributed, possessing a better thawing action and therefore reducing harmful effects on the environment. However, in order to achieve these advantages calcium chloride must be added to the salt brine in order to prevent it from freezing at only a few degrees below zero. Its corrosive action is encouraged by its hygroscopic character and the fact that the road surface dries more slowly, with the result that moist accumulations of dirt remain attached to the vehicles for a longer period. Fast luxury cars used mainly for long journeys even in winter on German "autobahns" and similar highways which are wet with salt are exposed to a particularly severe burden in this respect.

Stone chips are being adopted increasingly for spreading on roads where the risk of contaminating the ground water with chemical substances is particularly high. The number of local authorities which spread only stone chips on roads in their areas for reasons of general environmental protection is rising all the time.

One can therefore sum up by saying that the average amount of rusting which occurs on road vehicles is largely dependent on the area in which they operate and on the amount of initial mechanical damage they suffer in the form of penetration of the coatings applied to the sheet metal when struck by stones or chips thrown up from the road surface. A dominant stress factor in the electrochemical corrosion process is the amount of salt (electrolyte) spread on specific roads each year.

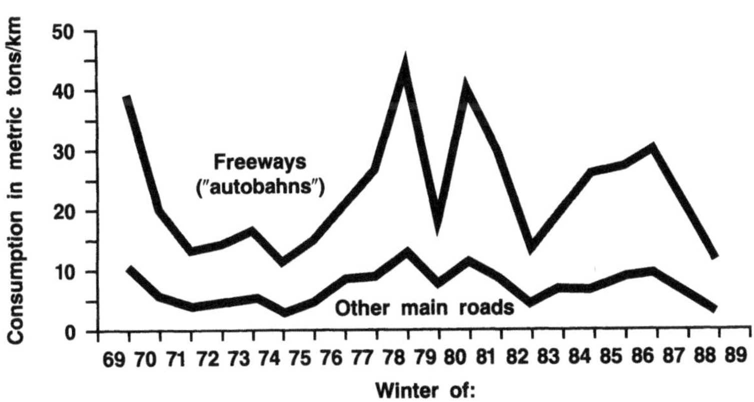

Fig. 2
Average consumption of road salt in the Federal Republic of Germany [2, 3]

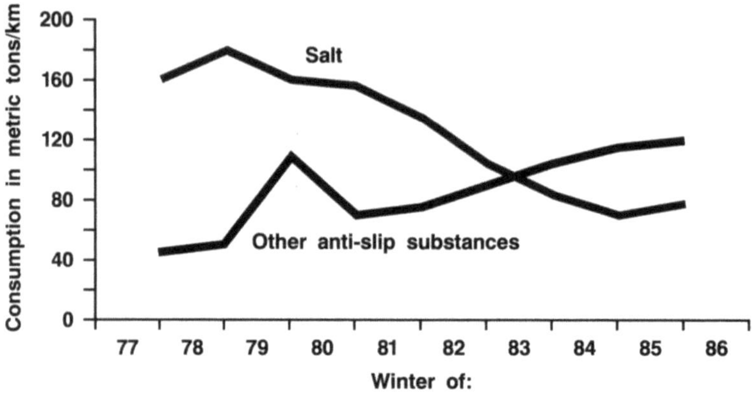

Fig. 3
Consumption of salt and other anti-slip substances by public authorities [4]

Any vehicle corrosion testing procedure must therefore simulate these conditions in such a way that the resulting corrosion pattern is as authentic as possible.

3.3 Corrosion on automobiles in the Federal Republic of Germany

What are the characteristic forms and areas of corrosion on automobiles mostly used in the Federal Republic of Germany? A reader survey conducted by the "Stiftung Warentest" consumer organization established that according to statements made by car owners the most severe rusting on the "average car" occurred on doors and door sills.

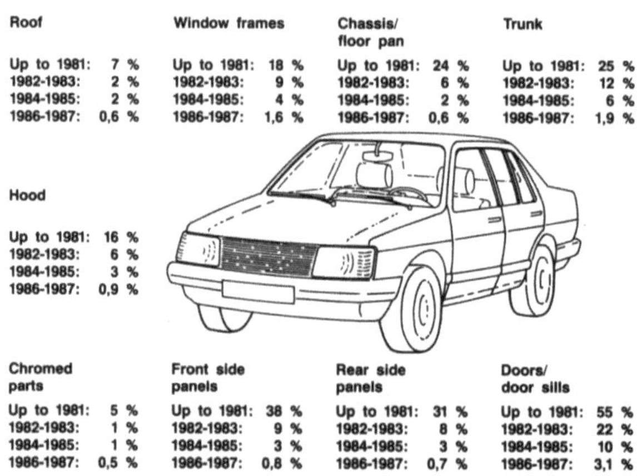

Fig. 4
How often the average car rusts [5]

The next chart shows the most frequent defects detected during regular official inspections conducted in 1987 by the Rhine-Westphalian Technical Inspection Authority (TÜV). In the case of vehicles eight years old or older, corrosion was the most frequent cause of complaint.

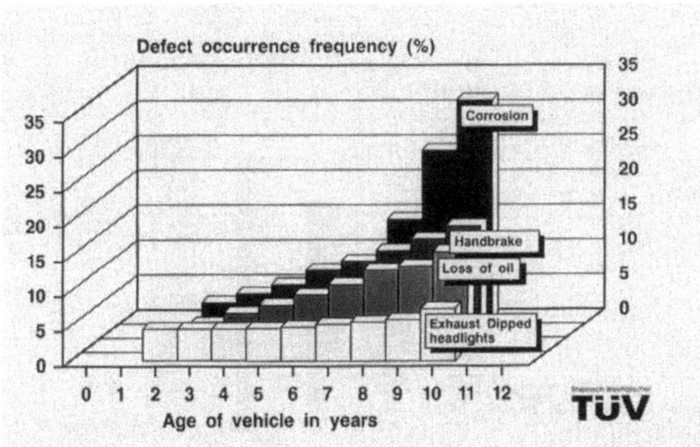

Fig. 5
The most frequent defects detected by official inspections (1987) [6]

Compared with the frequency of defects on brake lines and exhaust systems caused by corrosion, body corrosion defects increase considerably in significance from the seventh year of ownership onwards.

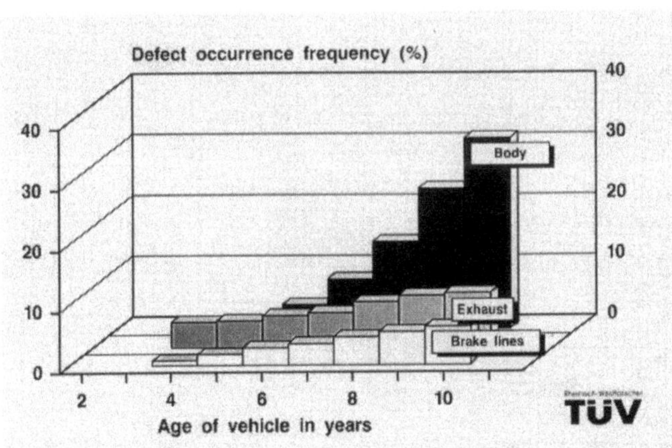

Fig. 6
Frequency of other corrosion defects compared with vehicle body (1987) [6]

When the distribution of corrosion among the various parts of the vehicle's structure is examined, it is found to confirm the results of earlier investigations, namely that the outer panels of the body exhibit the most severe corrosion propagation after only four years of operation.

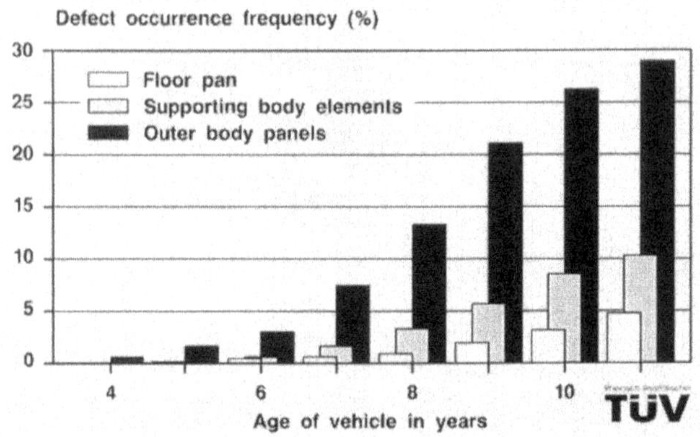

Fig. 7
Corrosion of various parts of the vehicle body (1985) [6]

Once again, corrosion of cavities in supporting structural elements is of particular importance, since the high complexity and cost of repair work quite often leads to older vehicles having to be taken off the road.

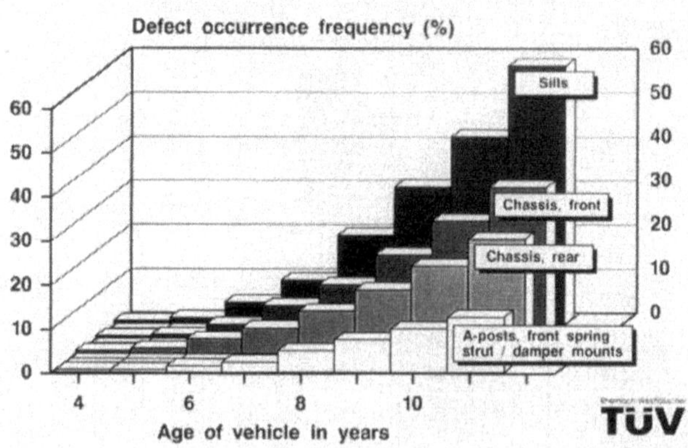

Fig. 8
Cavity corrosion (1986) [6]

These observations enable us to draw the following conclusions concerning the serviceability and resale value of older vehicles:

1. The survey of vehicle owners conducted by "Stiftung Warentest" revealed that 22 % of those which had been owned for five or six years and had therefore been manufactured in 1982 or 1983 were suffering from door and door sill rusting.
2. According to the German Technical Inspection Authority's own records, the most frequent points at which cavity corrosion in supporting structural elements of the body occurs are the door sills, where partial rusting through is normally encountered. With a defect frequency of approx. 8 % in the sixth year of ownership, this is accordingly the dominant risk factor in the long-term quality of the vehicle.
3. Easily the most frequent cases of corrosion complained of by the German "TÜV" are those concerning the outer body panels, starting as early as the fourth year of ownership.

This is a surprising result if we consider that when the vehicles concerned were produced every manufacturer in the Federal Republic of Germany had already studied the problem of corrosion protection most intensively and possessed the appropriate testing facilities. The consequence can only be that manufacturers are called upon to take further steps towards "harmonizing" the motor vehicle ageing process, that is to say ageing of the body, for instance, must not take place at a significantly more rapid rate than wear in the engine, transmission or other complex systems.

4. CORROSION TEST OF COMPLETE AUTOMOBILES AT BMW

We must therefore ask ourselves what these conclusions mean when determining a suitable corrosion testing procedure.

4.1. BMW's objectives

The basic philosophy of the Bayerische Motoren Werke AG as the manufacturers of BMW automobiles is for these to attain a peak quality standard which complies with customers' expectations in the market segment the company's products occupy. This applies equally to corrosion testing standards. Excellent resistance to corrosion must be assured by the availability of reliable test results, as a contribution towards overall vehicle quality.

The duration and severity of the loads imposed during the test procedure had therefore to be chosen to provide reliable confirmation of the avoidance not only of the "cosmetic" body surface rusting, to which customers strongly object, but also of rust penetration within or after the relevant warranty period. A particular requirement was accurate simulation of the corrosive action occurring in the cavities of the supporting body structure. These target requirements for the presentation of results from the corrosion testing procedure led to the definition of the key stages in the main simulation processes.

4.2. The DYKO 3 dynamic corrosion testing procedure

DYKO 3 was developed from the preceding accelerated corrosion testing procedures for complete automobiles known as DYKO 1 and 2, making use of the new corrosion testing site built for the purpose.

It encompasses the following characteristic loads:

1. Stationary exposure to salt spray mist in alternating cycles or combined with high atmospheric humidity and high reaction temperature (an aggressive atmosphere)
2. Dynamic exposure during actual driving, with the vehicle bombarded by stone chips, sand and gravel either thrown up by its own wheels or those of other vehicles, to cause initial damage and erosion of the sheet-metal coatings
3. Bringing the vehicle into contact with salt slurry, salt water and salt water spray in conditions closely resembling those actually encountered by customers.

The individual test processes have been optimized according to the overall objective of "maximum corrosion pattern authenticity". Shortcomings in this respect due to the accelerated time scale have been deliberately taken into account in order to avoid results which are excessive or difficult to evaluate.

The DYKO 3 corrosion testing procedure takes place in the following stages:

- preparation of vehicle
- pre-conditioning
- 60 two-day test cycles
- evaluation

4.2.1. Preparation of the vehicle

The following preparatory work is of particular importance in order to ensure that the corrosion pattern can be accurately evaluated at the end of the test:

1. Recording the actual level of corrosion proofing measures applied to the vehicle (sealing, cavity protection, thickness of paint coatings etc.) and comparing them with specifications
2. Deliberate preliminary mechanical damage in the form of scratches
3. Attachment of reference panels to the front bumper
4. If necessary, installation of experimental components

4.2.2. Pre-conditioning

Using a standardized procedure, the test vehicles are subjected to initial loads and indeed to actual damage in extreme peripheral conditions with a view to obtaining early evidence by virtue of the accelerated test time scale of weak points in the corrosion protection treatment they have received.

When exposure has been completed a one-day drying phase follows before the actual corrosion test commences.

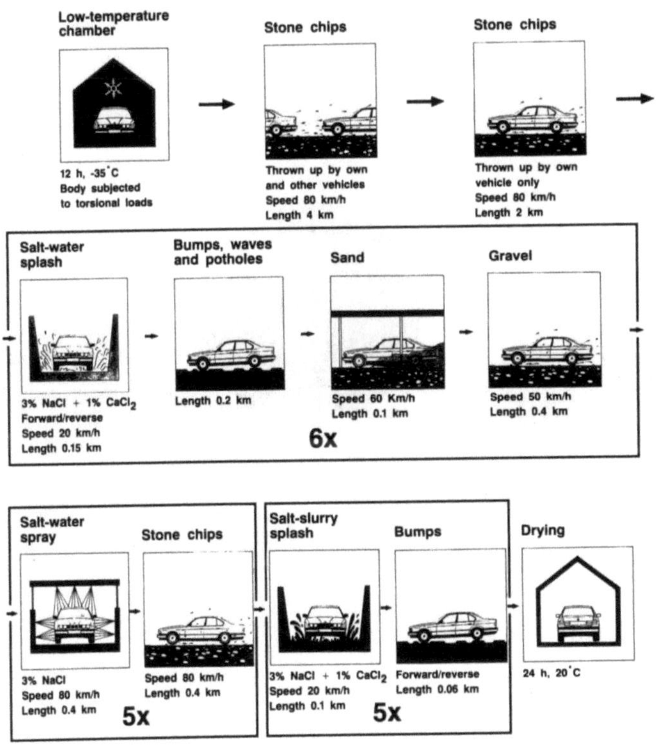

Fig. 9
BMW DYKO 3 corrosion test procedure - pre-conditioning

4.2.3. The two-day cycle

The two-day cycle consists of four phases. In phases 1 and 2, the vehicle remains in the climatic chamber throughout the first 24 hours (salt spray test according to German Industrial Standard DIN 50021 and condensation water test according to DIN 50017), in other words the stationary part of the test is performed.

On the second day, in phase 3, the vehicle is given a mobile or "dynamic" test. This mainly consists of passing through the various corrosion test site areas twice, with a preceding 90 kilometre drive to the test site on the German "autobahn" and a return drive to experimental department headquarters on ordinary main roads (110 km) conducted in actual traffic.

Phase 4, the incubation phase, enables workshop personnel and the responsible test engineers to carry out any necessary repairs and, for instance, to ensure that the doors, lids and windows can still be operated.

The test comprises a total of 60 cycles.

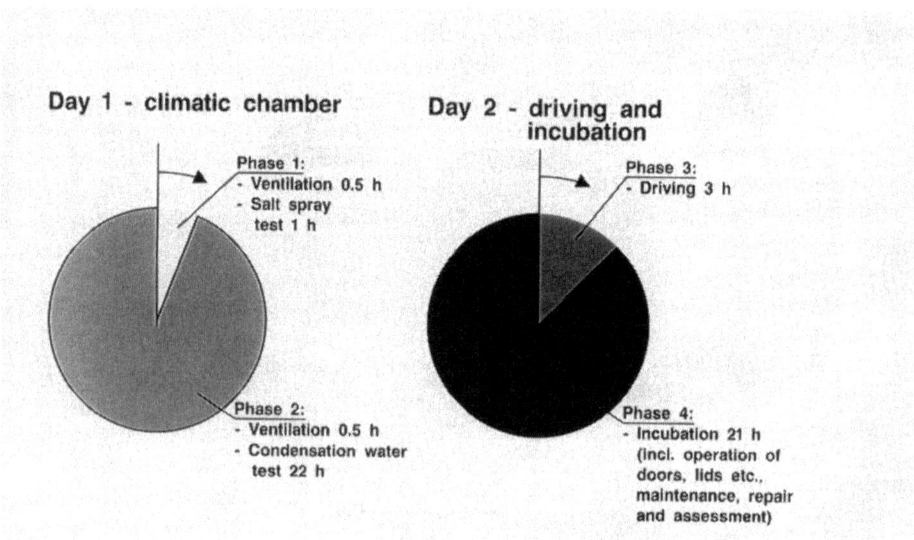

Fig. 10
Phases of two-day cycle

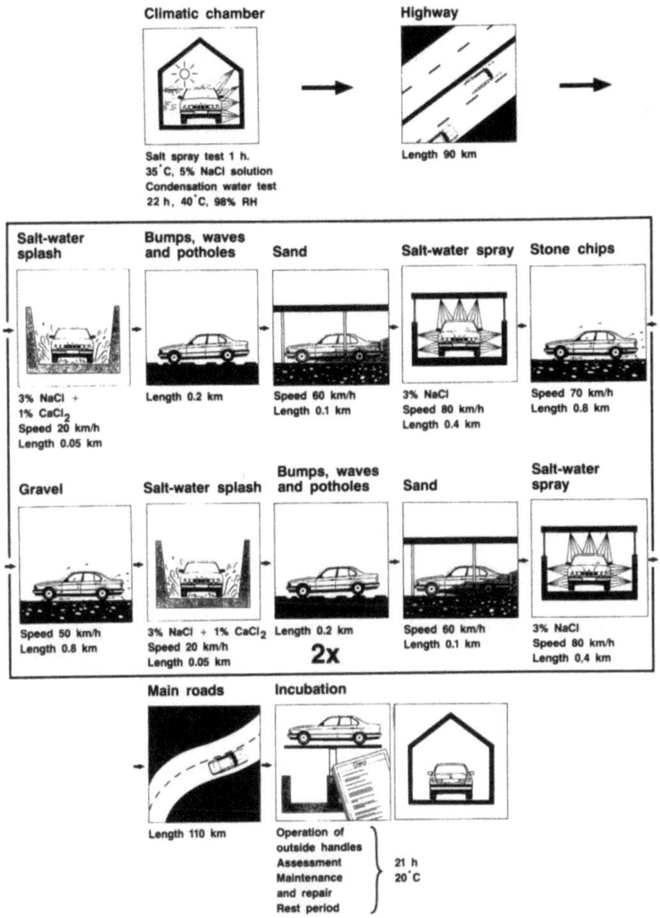

Fig. 11
BMW DYKO 3 corrosion test procedure: 2-day cycle

4.2.4. The testing facilities

In addition to the objectives laid down for the corrosion testing procedure, the following **technical design criteria** had to be taken into consideration when planning the testing facilities:

- suitability for long-term use, by the adoption of advanced technologies and structures and provision for adaptability (variations in testing methods)
- maximum cost-effectiveness by making optimum use of the available space, automatic process control with a high level of reliability and multi-purpose utilization of the installed facilities
- optimum reproducibility of the results
- minimum environmental burden in the surrounding area

The **climatic chambers** are of a combined pattern in which both salt spray tests and climatic simulation (for example condensation water tests) can be carried out.

This ensures that optimum use is made of the site area and the amount of time available. Phase 1 and 2 processes, that is to say the first 24 hours of the two-day cycle, take place fully automatically under program control.

The **corrosion test site** is within the boundaries of BMW's plant in Dingolfing, Bavaria, and has been in operation since February 1988. The road-test sections and the location of the buildings were chosen to ensure that maximum use was made of available space. The site measures 500 x 50 metres.

Fig. 12
General view of corrosion-test site

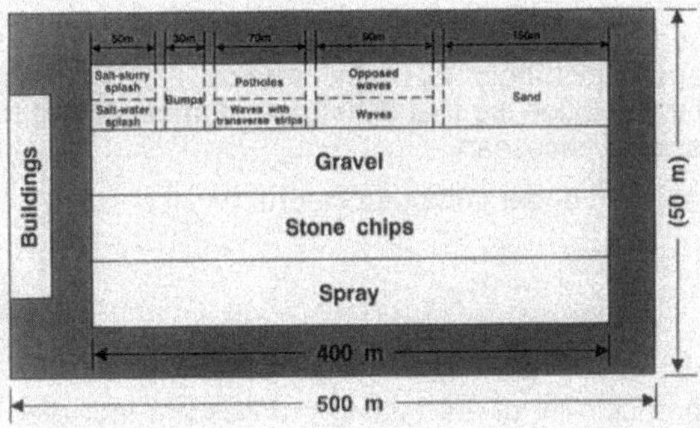

Fig. 13
Layout of test sections on corrosion-test site

The **spray section** is the heart of the test site. It simulates the passage of the car along an "autobahn" wet with salt in dense traffic (salt spray in the air). It takes the form of a tunnel 400 metres long with narrow windows at the top of the walls; speeds of up to 100 km/h can be reached inside the tunnel, so that the airflow over the vehicle is realistic.

Fig. 14
Spray section

Fig. 15
Salt solution recovery and storage

180 spray jets are distributed uniformly along the sides of the tunnel, in four separately controlled groups each 100 metres long. There are also 22 jets spread out along a 50 metre long section in the centre of the tunnel roof, which can be switched on if needed. A sequential control system ensures that the jets are only switched on as the test vehicle approaches; this keeps consumption of the 3 % NaCl solution to a minimum (approx. 1.1 cubic metre per test run).

After the salt solution has drained off it is collected in a vat and returned to the treatment plant. The two storage tanks hold 20 cubic metres and the actual treatment tank a further 10 m³ of the solution, so that a sufficient number of test runs can be performed in rapid succession if necessary.

The entire process is fully automatic and under computer control.

4.3. Evaluating the results

4.3.1. Methods of evaluation

The **degree of load** and the reproducibility of the testing process are indicated by the **amount of material eroded** from the **reference panels** attached at the front of the car (size 100 x 100 mm).

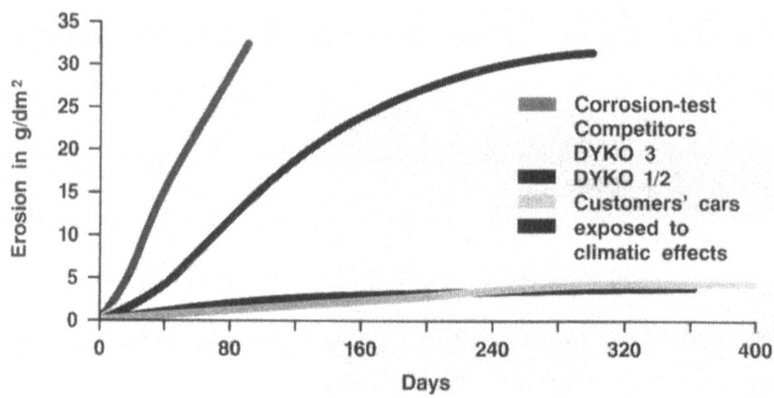

Fig. 16
Material erosion from reference panels as an indicator of test load

As a means of determining the extent to which **rust penetrates** beneath the paint coating on lids, doors, side panels and roof, **scratches** (length 100 mm, width 1.0 mm) are made before the start of the test to expose the bare sheet metal. The average **rust penetration width** is measured according to German VDA Test Sheet 621-415.

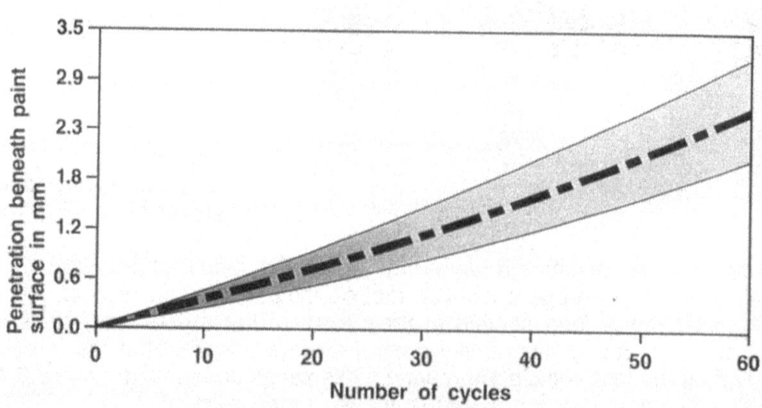

Fig. 17
Average penetration beneath scratched paint surface

4.3.2 Empirical assessment

Individual, empirical assessment of the results is by applying the following criteria

- number
- position (visible area)
- size and width of penetration
- material erosion (residual sheet-metal thickness)

to the corrosion surfaces, the assessment principles being different for "cosmetic" rusting and for body cavity corrosion. The extensive interim assessment undertaken after 30 cycles, including partial dismantling of the body, is mainly aimed at assessing cosmetic corrosion. Final assessment of cavity corrosion is only possible on completion of the 60-cycle test, since it requires the body to be dismantled and the various cavities opened up.

Comparability of results is achieved by operating vehicles **in pairs** in all cases, that is to say a second test vehicle with known corrosion characteristics is given precisely the same treatment as the test specimen itself. This ensures that largely reliable statements can be made regarding the degree of improvement achieved by a modified production process, a new material or a design change (a new model). In accordance with the usual practice, BMW also investigates competitors' products on occasion for purposes of comparison, if interesting results seem likely.

Correlation with the loads to which customers' cars are subjected in actual use is ensured by two different methods:

1. The systematic repurchase of cars from customers according to chassis number (manufacturing batch), operating region and nature of operation. This method has demonstrated the correlation between **DYKO 3 and an actual service load of approx. 6 years** in operating conditions which encourage severe corrosion.

2. Comparison with experimental vehicles in the **long-term test fleet**. These are BMW cars operated within the company by specific departments which use them for different but characteristic purposes. By operating these vehicles systematically it is possible to simulate a broad range of uses typical of those to which the customer puts the vehicle.

Discussions "at the vehicle" attended by representatives of top management always take place before modification decisions are taken or when results have to be evaluated. The vehicles are stripped down after the corrosion test and the results fully documented in test reports, including extensive photographs.

It should perhaps be stressed at this point that very considerable importance is attached to the experienced test engineers' subjective assessments during the process of achieving a reliable interpretation of the results.

Bibliography

[1] Kawamoto, J. and Nishii, K.:
The development of evaluation test procedures for vehicle corrosion caused by de-icing salts.

[2] Speth, O.:
Salzverbrauch beim Winterstrassendienst in Abhängigkeit von der Winterintensität (Salt consumption on winter roads in relation to the intensity of the climate during the winter season).
From "Strasse und Autobahn", No. 2/1988.

[3] Information issued by Federal German Ministry of Transport, August 1989.

[4] Durth, W.:
Stand und Entwicklung des Winterdienstes in Städten und Gemeinden (Current position and trends in winter road services in towns and local authority areas).
From "Strasse und Autobahn", No. 1/1989.

[5] "test" magazine, No. 9/89, publ. Stiftung Warentest.

[6] Grohnert, T.:
Häufigkeit von Korrosionsschäden an Pkw-Karosserien (Frequency of corrosion damage on automobile bodies).
From the "Praxis-Forum" series, in "Oberflächentechnik" No. 10/89

This paper was originally published by Vulkan-Verlag, Essen, Germany for "Haus der Technik" in a collection entitled "Korrosionsschutz bei Kraftfahrzeugen" (Motor Vehicle Corrosion Protection).

MIX
Papier aus verantwortungsvollen Quellen
Paper from responsible sources
FSC® C105338

If you have any concerns about our products,
you can contact us on
ProductSafety@springernature.com

In case Publisher is established outside the EU,
the EU authorized representative is:
Springer Nature Customer Service Center GmbH
Europaplatz 3, 69115 Heidelberg, Germany

Printed by Libri Plureos GmbH
in Hamburg, Germany